T0176974

IMAGE, VIDEO & 3D DATA REGISTRATION

IMAGE, VIDEO & 3D DATA REGISTRATION

MEDICAL, SATELLITE & VIDEO PROCESSING APPLICATIONS WITH QUALITY METRICS

Vasileios Argyriou

Kingston University, UK

Jesús Martínez del Rincón

Queen's University Belfast, UK

Barbara Villarini

University College London, UK

Alexis Roche

Siemens Healthcare / University Hospital Lausanne / École Polytechnique Fédérale Lausanne, Switzerland

WILEY

This edition first published 2015

© 2015 John Wiley & Sons Ltd

Registered office
John Wiley & Sons Ltd, The Atrium, Southern Gate, Chichester, West Sussex, PO19 8SQ, United Kingdom

For details of our global editorial offices, for customer services and for information about how to apply for permission to reuse the copyright material in this book please see our website at www.wiley.com.

The right of the author to be identified as the author of this work has been asserted in accordance with the Copyright, Designs and Patents Act 1988.

All rights reserved. No part of this publication may be reproduced, stored in a retrieval system, or transmitted, in any form or by any means, electronic, mechanical, photocopying, recording or otherwise, except as permitted by the UK Copyright, Designs and Patents Act 1988, without the prior permission of the publisher.

Wiley also publishes its books in a variety of electronic formats. Some content that appears in print may not be available in electronic books.

Designations used by companies to distinguish their products are often claimed as trademarks. All brand names and product names used in this book are trade names, service marks, trademarks or registered trademarks of their respective owners. The publisher is not associated with any product or vendor mentioned in this book.

Limit of Liability/Disclaimer of Warranty: While the publisher and author have used their best efforts in preparing this book, they make no representations or warranties with respect to the accuracy or completeness of the contents of this book and specifically disclaim any implied warranties of merchantability or fitness for a particular purpose. It is sold on the understanding that the publisher is not engaged in rendering professional services and neither the publisher nor the author shall be liable for damages arising herefrom. If professional advice or other expert assistance is required, the services of a competent professional should be sought.

Library of Congress Cataloging-in-Publication Data

Image, video & 3D data registration : medical, satellite and video processing applications with quality metrics / [contributions by] Vasileios Argyriou, Faculty of Science, Engineering and Computing, Kingston University, UK, Jesús Martínez del Rincón, Queen's University, Belfast, UK, Barbara Villarini, University College London, UK, Alexis Roche, Siemens Medical Solutions, Switzerland.
        pages cm
    Includes bibliographical references and index.
    ISBN 978-1-118-70246-8 (hardback)
    1. Image registration. 2. Three-dimensional imaging. I. Argyriou, Vasileios. II. Title: Image, video and 3D data registration.
    TA1632.I497 2015
    006.6′93−dc23

                                   2015015478

A catalogue record for this book is available from the British Library.

Cover Image: Courtesy of Getty images

Set in 11/13pt, TimesLTStd by SPi Global, Chennai, India
Printed and bound in Singapore by Markono Print Media Pte Ltd

1   2015

Contents

Preface

This book was motivated by the desire we and others have had to further the evolution of the research in computer vision and video processing, focusing on image and video techniques for registration and quality performance metrics. There are a significant number of registration methods operating at different levels and domains (e.g. block or feature based, pixel level, Fourier domain), each method applicable in specific domains. Image registration or motion estimation in general is the process of calculating the motion of a camera and/or the motion of the individual objects composing a scene. Registration is essential for many applications such as video coding, tracking, object and face detection and recognition, surveillance and satellite imaging, structure from motion, simultaneous localization and mapping, medical image analysis, activity recognition for entertainment, behaviour analysis and video restoration.

In this book, we present the state-of-the-art registration based on the targeted application providing an introduction to the particular problems and limitations of each domain, an overview of the previous approaches and a detailed analysis of the most well-known current methodologies. Additionally, various assessment metrics for measuring the quality of registration are presented showcasing the differences among different targeted applications. For example, the important features in a medical image (e.g. MRI data) may be different from a human face picture, and therefore, the quality metrics are adjusted accordingly. The state-of-the-art metrics for quality assessment is analysed explaining their advantages and disadvantages and providing visual examples. Also information about common datasets utilized to evaluate these approaches is discussed for each application.

The evolution of the research related to registration and quality metrics has been significant over recent decades with popular examples including simple block matching techniques, optical flow and feature-based approaches. Also, over the last few years with the advent of hardware architectures, real-time algorithms have been introduced. In the near future, it is expected to have high-resolution images and videos processed in real time. Furthermore, the advent of new acquisition devices capturing new modalities such as depth require traditional concepts in registration and the quality assessment to be revised while new applications are also discussed.

This book will provide:

- an analysis of registration methodologies and quality metrics covering the most important research areas and applications;
- an introduction to key research areas and the current work underway in these areas;
- to an expert on a particular area the opportunity to learn about approaches in different registration applications and either obtain ideas from them, or apply his or her expertises to a new area improving the current approaches and introducing novel methodologies;
- to new researchers an introduction up to an advanced level, and to specialists, ways to obtain or transfer ideas from different areas covered in this book.

Acknowledgements

We are deeply indebted to many of our colleagues who have given us valuable suggestions for improving the book. We acknowledge helpful advice from Professor Theo Vlachos and Dr. George Tzimiropoulos during the preparation of some of the chapters.

1

Introduction

In the last few decades, the evolution in technology has provided a rapid development in image acquisition and processing, leading to a growing interest in related research topics and applications including image registration. Registration is defined as the estimation of a geometrical transformation that aligns points from one viewpoint of a scene with the corresponding points in the other viewpoint. Registration is essential in many applications such as video coding, tracking, detection and recognition of object and face, surveillance and satellite imaging, structure from motion, simultaneous localization and mapping, medical image analysis, activity recognition for entertainment, behaviour analysis and video restoration. It is considered one of the most complex and challenging problems in image analysis with no single registration algorithm to be suitable for all the related applications due to the extreme diversity and variety of scenes and scenarios. This book presents image, video and 3D data registration techniques for different applications discussing also the related quality performance metrics and datasets. State-of-the-art registration methods based on the targeted application are analysed, including an introduction to the problems and limitations of each method. Additionally, various assessment quality metrics for registration are presented indicating the differences among the related research areas. For example, the important features in a medical image (e.g. MRI data) may not be the same as in the picture of a human face, and therefore the quality metrics are adjusted accordingly. Therefore, state-of-the-art metrics for quality assessment are analysed explaining their advantages and disadvantages, and providing visual examples separately for each of the considered application areas.

1.1 The History of Image Registration

In image processing, one of the first times that the concept of registration appeared was in Roberts' work in 1963 [1]. He located and recognized predefined polyhedral objects

Image, Video & 3D Data Registration: Medical, Satellite & Video Processing Applications with Quality Metrics,
First Edition. Vasileios Argyriou, Jesus Martinez Del Rincon, Barbara Villarini and Alexis Roche.
© 2015 John Wiley & Sons, Ltd. Published 2015 by John Wiley & Sons, Ltd.

in scenes by aligning their edge projections with image projections. The first registration applied to an image was in the remote sensing literature. Using sum of absolute differences as similarity measure, Barnea and Silverman [2] and Anuta [3, 4] proposed some automatic methods to register satellite images. In the same years, Leese [5] and Pratt [6] proposed a similar approach using the cross-correlation coefficient as similarity measure. In the early 1980s, image registration was used in biomedical image analysis using data acquired from different scanners measuring anatomy. In 1973, for the first time Fischler and Elschlager [7] used non-rigid registration to locate deformable objects in images. Also, non-rigid registration was used to align deformed images and to recognize handwritten letters. In medical imaging, registration was employed to aligned magnetic resonance (MR) and computer tomography (CT) brain images trying to build an atlas [8, 9].

Over the last few years due to the advent of powerful and low-cost hardware, real-time registration algorithms have been introduced, improving significantly their performance and accuracy. Consequently, novel quality metrics were introduced to allow unbiased comparative studies. This book will provide an analysis of the most important registration methodologies and quality metrics, covering the most important research areas and applications. Through this book, all the registration approaches in different applications will be presented allowing the reader to get ideas supporting knowledge transfer from one application area to another.

1.2 Definition of Registration

During the last decades, automatic image registration became essential in many image processing applications due to the significant amount of acquired data. With the term *image registration*, we define the process of overlaying two or more images of the same scene captured in different times and viewpoints or sensors. It represents a geometrical transformation that aligns points of an object observed from a viewpoint with the corresponding points of the same or different object captured from another viewpoint. Image registration is an important part of many image processing tasks that require information and data captured from different sources, such as image fusion, change detection and multichannel image restoration. Image registration techniques are used in different contexts and types of applications. Typically, it is widely used in computer vision (e.g. target localization, automatic quality control), in remote sensing (e.g. monitoring of the environment, change detection, multispectral classification, image mosaicing, geographic systems, super-resolution), in medicine (e.g. combining CT or ultrasound with MR data in order to get more information, monitor the growth of tumours, verify or improve the treatments) and in cartography updating maps. Image registration is also employed in video coding in order to exploit the temporal relationship between successive frames (i.e. motion estimation techniques are used to remove temporal redundancy improving video compression and transmission).

In general, registration techniques can be divided into four main groups based on how the data have been acquired [10]:

- *Different viewpoints (multiview analysis)*: A scene is acquired from different viewpoints in order to obtain a larger/panoramic 2D view or a 3D representation of the observed scene.
- *Different times (multitemporal analysis)*: A scene is acquired in different times, usually on a regular basis, under different conditions, in order to evaluate changes among consecutive acquisitions.
- *Different sensors (multimodal analysis)*: A scene is acquired using different kinds of sensors. The aim is to integrate the information from different sources in order to reveal additional information and complex details of the scene.
- *Scene to model registration*: The image and the model of a scene are registered. The model can be a computer representation of the given scene, and the aim is to locate the acquired scene in the model or compare them.

It is not possible to define a universal method that can be applied to all registration tasks due to the diversity of the images and the different types of degradation and acquisition sources. Every method should take different aspects into account. However, in most of the cases, the registration methods consist of the following steps:

- *Feature detection*: Salient objects, such as close-boundary regions, edges, corners, lines and intersections, are manually or automatically detected. These features can be represented using points such as centre of gravity and line endings, which are called control points (CPs).
- *Feature matching*: The correspondence between the detected features and the reference features is estimated. In order to establish the matching, features, descriptors and similarity measures among spatial relationships are used.
- *Transform model estimation*: According to the matched features, parameters of mapping functions are computed. These parameters are used to align the sensed image with the reference image.
- *Image resampling and transformation*: The sensed image is transformed using the mapping functions. Appropriate interpolation techniques can be used in order to calculate image values in non-integer coordinates.

1.3   What is Motion Estimation

Video processing differs from image processing due to the fact that most of the observed objects in the scene are not static. Understanding how objects move helps to transmit, store and manipulate video in an efficient way. Motion estimation is the research area of imaging a video processing that deals with these problems, and it is also linked to feature matching stage of the registration algorithms. Motion

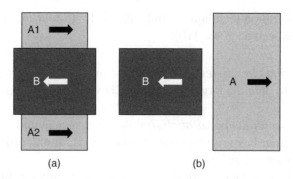

Figure 1.1   (a) Occluded objects A1 and A2, (b) single object A

estimation is the process by which the temporal relationship between two successive frames in a video sequence is determined. Motion estimation is a registration method used in video coding and other applications to exploit redundancy mainly in the temporal domain.

When an object in a 3D environment moves, the luminance of its projection in 2D is changing either due to non-uniform lighting or due to motion. Assuming uniform lighting, the changes can only be interpreted as movement. Under this assumption, the aim of motion estimation techniques is to accurately model the motion field. An efficient method can produce more accurate motion vectors, resulting in the removal of a higher degree of correlation.

Integer pixel registration may be adequate in many applications, but some problems require sub-pixel accuracy, either to improve the compression ratio or to provide a more precise representation of the actual scene motion. Despite the fact that sub-pixel motion estimation requires additional computational power and execution time, the obtained advantages settle its use that is essential for the most multimedia applications.

In a typical video sequence, there is no 3D information about the scene contents. The 2D projection approximating a 3D scene is known as 'homography', and the velocity of the 3D objects corresponds to the velocity of the luminance intensity on the 2D projection, known as 'optical flow'. Another term is 'motion field', a 2D matrix of motion vectors, corresponding to how each pixel or block of pixels moves. General 'motion field' is a set of motion vectors, and this term is related to the 'optical flow' term, with the latter being used to describe dense 'motion fields'.

Finding the motion between two successive frames of a video sequence is an ill-posed problem due to the intensity variations not exactly matching the motion of the objects. Another problematic phenomenon is the covered objects, in which case it is efficient to make the assumption that the occluded objects can be considered as many separable objects, until they are observed as a single object (Figure 1.1). Additionally, in motion estimation, it is assumed that motion within an object is smooth and uniform due to the spatial correlation.

The concept of motion estimation is used in many applications and is analysed in the following chapters providing details of state-of-the-art algorithms, allowing the reader to apply this information in different contexts.

1.4 Video Quality Assessment

The main target in the design of modern multimedia systems is to improve the video quality perceived by the user. Video quality assessment is a difficult task because many factors can interfere on the final result.

In order to obtain quality improvement, the availability of an objective quality metric that represents well the human perception is crucial. Many methods and measures have been proposed aiming to provide objective criteria that give accurate and repeatable results taking into account the subjective experience of a human observer. Objective quality assessment methods based on subjective measurements are using either a perceptual model of the human visual system (HVS) or a combination of relevant parameters tuned with subjective tests [11, 12].

Objective measurements are used in many image and video processing applications since they are easy to apply for comparative studies. One of the most popular metrics is peak signal-to-noise ratio (PSNR) that is based on the mean square error between the original and a distorted data. The computation of this value is trivial but has significant limitations. For example, it does not correlate well with the perceived quality, and in many cases the original undistorted data (e.g. images, videos) may not be available.

At the end of each chapter, a description of the metrics used to assess the quality of the presented registration methods is available for all the discussed applications, highlighting the key factors that affect the overall quality, the related problems and solutions, and the examples to illustrate these concepts.

1.5 Applications

1.5.1 Video Processing

Registration techniques are required in many applications based on video processing. As mentioned in the earlier section, motion estimation is a registration task employed to determine the temporal relationship between the video frames. One of the most important applications of motion estimation is in video coding systems.

Video CODECs (COder/DECoder) comprise an encoder and a decoder. The encoder compresses (encodes) video data resulting in a file that can be stored or streamed economically. The decoder decompresses (decodes) encoded video data (whether from a stored file or streamed), enabling video playback.

Compression is a reversible conversion of data to a format that requires fewer bits, usually performed so that the data can be stored or transmitted more efficiently. The size of the data in compressed form C relative to the original size O is known as the

Figure 1.2   Motion estimation predicts the contents of each macroblock base due the motion relative to the reference frame. The reference frame is searched to find the 16×16 block that matches the macroblock

compression ratio $R = O/C$. If the inverse of the process, 'decompression', produces an exact replica of the original data, then the compression is lossless. Lossy compression, usually applied to image and video data, does not allow reproduction of an exact replica of the original data but results in higher compression ratios.

Neighbouring pixels within an image or a video frame are highly correlated (spatial redundancy). Also neighbouring areas within successive video frames are highly correlated too (temporal redundancy).

A video signal consists of a sequence of images. Each image can be compressed individually without using the other video frames (intra-frame coding) or can exploit the temporal redundancy considering the similarity among consecutive frames (inter-frame coding), obtaining a better performance. This is achieved in two steps:

1. *Motion estimation*: A region (usually a block) of the current frame is compared with neighbouring region of the adjacent frames. The aim is to find the best match typically in the form of motion vectors (Figure 1.2).
2. *Motion compensation*: The matching region from the reference frame is subtracted from the current region block.

Motion estimation considers images of the same scene acquired in different time, and for this reason it is regarded as an image registration task. In Chapter 2, the most popular motion estimation methods for video coding are presented.

Motion estimation is not only utilised in video coding applications but also to improve the resolution and the quality of the video. If we have multiple, shifted and low-resolution images, we can use image processing methods in order to obtain high-resolution images.

Furthermore, digital videos acquired by consumer camcorders or high-speed cameras, which can be used in industrial applications and to track high-speed objects, are often degraded by linear space-varying blur and additive noise. The aim of video restoration is to estimate each image or frame, as it would appear without the effects of sensor and optics degradations. Image and video/restoration are essential when we want to extract still images from videos. This is because blurring and noise may not be visible to the human eye at usual frame rates, but they can become rather evident when observing a 'freeze-frame'. The restoration is also a technique used when historical film materials are encoded in a digital format. Especially if they are encoded with block-based encoders, many artefacts may be present in the coded frame. These artefacts are removed using sophisticated techniques based on motion estimation. In Chapter 8, video registration techniques used in restoration applications are presented.

1.5.2   Medical Applications

Medical images are increasingly employed in health care for different kinds of tasks, such as diagnosis, planning, treatment, guided treatment and monitoring diseases progression. For all these studies, multiple images are acquired from subjects at different times and in the most of the cases using different imaging modalities and sensors. Especially with the growing number of imaging systems, different types of data are produced. In order to improve and gain information, proper integration of these data is highly desirable. Registration is then fundamental in this integration process. One example of different data registration is the epilepsy surgery. Usually the patients undergo various data acquisition processes including MR, CT, digital subtraction angiography (DSA), ictal and interictal single-photon emission computed tomography (SPECT) studies, magnetoencephalography (MEG), electroencephalography (EEG) and positron emission tomography (PET). Another example is the radiotherapy treatment, in which both CT and MR are employed. Therefore, it can be argued that the benefits for the surgeons are significant by registering all these data. Registering methods are also applied to monitor the growth of a tumour or to compare the patient's data with anatomical atlases.

Motion estimation is also used for medical applications operating like a doctor's assistant or guide. For example, motion estimation is used to indicate the right direction for the laser, displaying the optical flow (OF) during interstitial laser therapy (ILT) of a brain tumour. The predicted OF velocity vectors are superimposed on the grey-scaled images, and the vectors are used to predict the amount and the direction of heat deposition.

Another growing application of registration is in recognition of face, lips and feelings using motion estimation. A significant amount of effort has been put on sign language recognition. The motion and the position of the hand and the fingers are estimated, and patterns are used to recognize the words and the meanings (see Figure 1.3), an application particularly useful for deaf-mute people [13].

The main problems of medical image data analysis and the application of registration techniques are discussed in details in Chapter 7.

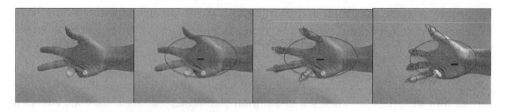

Figure 1.3   Steps of hand and fingers motion estimation

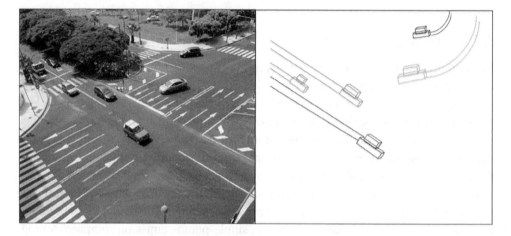

Figure 1.4   The output of a tracking system

1.5.3   Security Applications

Registration methods find many applications in systems used to increase the security for persons and vehicles. Video processing and motion estimation can be used to protect humans from both active and passive accidents. The tracking of vehicles has many potential applications, including road traffic monitoring (see Figure 1.4), digital rear-view mirror, monitoring of car parks and other high-risk or high-security sites. The benefits are equally wide ranging. Tracking cars in roads could make it easier to detect accidents, potentially cutting down the time it takes for the emergency services to arrive.

Another application of motion interpretation in image sequences is a driver assistance system for vehicles driving on the highway. The aim is to develop a digital rear-view mirror. This device could inform the driver when a lane-shift is unsafe. A single camera-based sensor can be used to retrieve information of the vehicle's environment in question. Vehicles driving behind the vehicle in question limit its motion possibilities [14]. Therefore, the application needs to estimate their motion relative to the first car. This problem is illustrated in Figure 1.5.

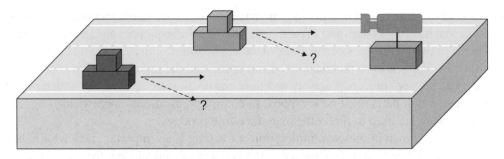

Figure 1.5 Observing other vehicles with a single camera

(a)   (b)

Figure 1.6 (a) Intelligent CCTV system, (b) 3D indoor visual system

Tracking human motion can be useful for security and demographic applications. For example, an intelligent CCTV system (see Figure 1.6(a)) will be able to 'see' and 'understand' the environment. In shopping centres, it will be able to count the customers and to provide useful information (e.g. number of customers on a particular day or in the last 2 h). Robots can be utilised for indoor applications without the need of prior knowledge of the building being able to move and perform specific operations. This kind of application can be used in hospitals or offices where the passages can be easily identified (see Figure 1.6(b)).

Visual tracking is an interesting area of computer vision with many practical applications. There are good reasons to track a wide variety of objects, including aeroplanes, missiles, vehicles, people, animals and microorganisms. While tracking single objects alone in images has received considerable attention, tracking multiple objects simultaneously is both more useful and more problematic. It is more useful since the objects to be tracked often exist in close proximity to other similar objects. It is more problematic since the objects of interest can touch, occlude and interact with each other; they can also enter and leave the scene.

The above-mentioned applications are based on machine vision technology, which utilises an imaging system and a computer to analyse the sequences and take decisions. There are two basic types of machine vision applications – inspection and control. In inspection applications, the machine vision optics and the imaging system enable the processor to 'see' objects precisely and thus make valid decisions about which parts pass and which parts must be scrapped. In control applications, sophisticated optics and software are used to direct the manufacturing process.

As it was shown in these examples, object tracking is an important task within the field of computer vision. In Chapter 3, the concept of optical flow for tracking and activity recognition are analysed presenting the related registration methodologies.

Considering a security system, another important application is face tracking and recognition. It is a biometric system for automatically identifying and verifying a person from a video sequence or frame. It is now very common to find security cameras in airports, offices, banks, ATMs and universities and in any place with an installed security system. Face recognition should be able to detect initially a face in an image. Then features are extracted that are used to recognize the human, taking into account factors such as lighting, expression, ageing, illumination, transformation and pose. Registration methods are used especially for face alignment tasks and they are presented in Chapter 4 highlighting the main problems, the different approaches and metrics for the evaluation of these tasks.

1.5.4 Military and Satellite Applications

Military applications are probably one of the largest areas for computer vision, even though only a small part of the work is open to the public. The obvious examples are detection of enemy soldiers or vehicles and guidance of missiles to a designated target. More advanced systems for missile guidance send the missile to an area rather than a specific target, and target selection is made when the missile reaches the area based on locally acquired image data. Modern military concepts, such as 'battlefield awareness', imply that various sensors, including image sensors, provide a rich set of information about a combat scene, which can be used to support strategic decisions. In this case, automatic data processing is used to reduce complexity and to fuse information from multiple sensors increasing reliability.

Night and heat vision can be used by police to hunt a criminal. Motion estimation techniques and object tracking systems use these vision systems to obtain better performance at night and in cold areas. Torpedoes, bombs and missiles use motion estimation to find and follow a target. Also, motion estimation is utilised by aeroplanes, ships and submarines along with object tracking systems.

Satellite object tracking is one of the military applications with the most research programmes. In this case, the systems track objects such as vehicles, trains, aeroplanes or ships. The main problem in these applications is the low quality of the pictures. Therefore, these practical problems can be reduced when efficient algorithms are used.

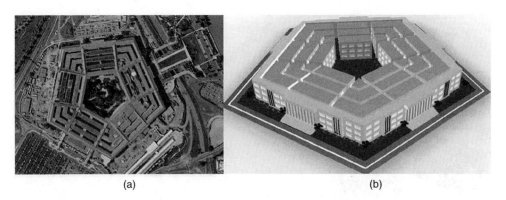

(a)   (b)

Figure 1.7   (a). Fragment of the satellite image, (b) 3D visualization of the part of the city, obtained as the result of the high-resolution satellite image processing. *Source:* http://commons .wikimedia.org/wiki/File:Pentagon-USGS-highres-cc.jpg

Satellite high-resolution images can also be used to produce maps and 3D visualization of territory status of a city (see Figure 1.7). In this case, image alignment is essential to obtain accurate maps and visualizations; therefore, sub-pixel registration methods are required. Also, due to the size of the captured data, fast approaches are crucial for this type of military and satellite applications. In Chapter 5, a detailed analysis of satellite image registration is presented. Methods for evaluating the performance of these techniques in the context of achieving specific performance goals are also described.

1.5.5   Reconstruction Applications

A fundamental task in computer vision is the one referred to as image-based 3D reconstruction for the creation of 3D models of a real scene from 2D images. Three-dimensional digital models are used in applications such as animation, visualization and navigation. One of the most common techniques for the image-based 3D reconstruction is structure from motion due to its conceptual simplicity. Structure from motion is a mechanism of simultaneously estimating the 3D geometry of a scene (structure) and the camera location (3D motion). Structure from motion is applied, for example, to reconstruct 3D archaeological building and statues using 2D images (see Figure 1.8).

In these approaches, the first step is to find the correspondence of sparse features among consecutive images using feature extraction and matching techniques. In the second step, structure from motion is applied in order to obtain the 3D shape and the motion from the camera. In the final step, both reconstruction and motion are adjusted and refined. Structure from motion is also used in different scenarios, such us autonomous navigation and guidance, augmented reality, hand/eyes motion capture, calibration, remote sensing and segmentation.

Figure 1.8   The 3D shape of a monument reconstructed using a sequence of 2D images

In Chapter 6, methods for obtaining the 3D shape of an object or an area using motion information and registration techniques are presented. Also registration methods for panoramic view for digital cameras and robotics are described focusing also on issues related to performance and computational complexity.

1.6   Organization of the Book

This book will provide an analysis on registration methodologies and quality metrics covering the most important research areas and applications. It is organized as follows:

Chapter 1 An introduction to the concepts of image and video registration is presented including examples of the related applications. An historical overview on image registration and the fundamentals on quality assessment metrics are also analysed.

Chapter 2 An overview of block-matching motion estimation methods is presented including traditional methods such as full search, tree steps, diamond and other state-of-the-art approaches. The same structure is used for hierarchical and shape-adaptive methods. The concepts of quality of system (QoS) and quality of experience (QoE) are discussed, and how image and video quality is influenced by the registration techniques is analysed. Quality metrics are presented focusing on coding applications.

Chapter 3 The concept of optical flow for tracking and activity recognition is analysed presenting the related state-of-the-art methods indicating their advantages and disadvantages. Approaches to extract the background and track objects and humans for surveillance are presented. Also, methods that operate in real time and focus on activity recognition both for security and entertainment (TV games) are discussed. The concepts of multitarget tracking evaluation including methods for detection recall, detection precision, number of ground truth detections, number of correct detections, number of false detections, number of total detections, and number of total tracks in ground truth are presented. Regarding action recognition metrics related to performance, time accuracy and quality of execution are also analysed.

Chapter 4 Registration methods for face alignment utilised mainly for recognition are presented. Both 2D and 3D techniques are included and different features are discussed. Problems that may affect the registration problem are overviewed including illumination and pose variations, occlusions and so on. Also, metrics for evaluation of these parts of the overall face recognition process are presented and analysed separately and as a whole.

Chapter 5 Methods for global motion estimation are introduced in this chapter operating both in the pixel and Fourier domains. The registration problem is further extended to rotation and scale estimation for satellite imaging applications. Methods for evaluating the performance of these techniques in the context of achieving specific performance are presented. Also the observed quality of registration methods is discussed.

Chapter 6 Methods for obtaining the 3D shape of an object or a scene using motion information and registration techniques are analysed. Additionally, registration methods for panoramic view for digital cameras and robotics are presented focusing on issues related to performance and computational complexity. Also the issues of performance and quality are addressed taking into consideration the related applications.

Chapter 7 The main problems of medical image data analysis are discussed. The particular characteristics of this type of data and the special requirements based on the exact application affect the registration methods. A variety of techniques are discussed applied on either 2D or 3D data. Also intra- and inter-subject image registration methods are reviewed such as non-rigid brain warping. Regarding the quality part, a similar structure with the other chapters is considered.

Chapter 8 Methods for video stabilization in real time based on motion estimation are presented. Furthermore, approaches to remove dirt, flickering and other effects during the restoration process of a video sequence based on registration are analysed, focusing on the latest approaches and the state-of-the-art techniques. Mechanisms for evaluating the restoration accuracy of algorithms operating in either real time or offline are discussed focusing on the quality of experience.

References

[1] Roberts, L.G. (1963) Machine perception of 3-D solids. Ph.D. Thesis. MIT.
[2] Barnea, D.I. and Silverman, H.F. (1972) A class of algorithms for fast digital image registration. *IEEE Transactions on Computers*, **21** (2), 179–186.
[3] Anuta, P.E. (1969) Registration of multispectral video imagery. *Society of Photographic Instrumentation Engineers Journal*, **7**, 168–175.
[4] Anuta, P.E. (1970) Spatial registration of multispectral and multitemporal digital imagery using fast Fourier transform techniques. *IEEE Transactions on Geoscience Electronics*, **8** (4), 353–368.
[5] Leese, J.A., Novak, G.S. and Clark, B.B. (1971) An automatic technique for obtaining cloud motion from geosynchronous satellite data using cross correlation. *Applied Meteorology*, **10**, 110–132.

[6] Pratt, W.K. (1974) Correlation techniques for image registration. *IEEE Transactions on Aerospace and Electronic Systems*, **10** (3), 353–358.

[7] Fischler, M.A. and Elschlager, R.A. (1973) The representation and matching of pictorial structures. *IEEE Transactions on Computers*, **22**, 67–92.

[8] Bajcsy, R., Lieberson, R. and Reivich, M. (1983) A computerized system for the elastic matching of deformed radiographic images to idealized atlas images. *Journal of Computer Assisted Tomography*, **7** (4), 618–625.

[9] Bohm, C., Greitz, T., Berggren, B.M. and Olsson, L. (1983) Adjustable computerized stereotaxic brain atlas for transmission and emission tomography. *American Journal of Neuroradiology*, **4**, 731–733.

[10] Zitova, B. and Flusser, J. (2003) Image registration methods: a survey. *Image and Video Computing*, **21**, 977–1000.

[11] Wang, Z., Bovik, A., Sheikh, H. and Simoncelli, E. (2004) Image quality assessment: from error measurement to structural similarity. *IEEE Transactions on Image Processing*, **13** (4), 600–612.

[12] Sheikh, R., Sabir, M. and Bovik, A.C. (2006) A statistical evaluation of recent full reference image quality assessment algorithms. *IEEE Transactions on Image Processing*, **15** (11), 3440–3451.

[13] Bretzner, L. and Lindeberg, T. (2000) Qualitative multi-scale feature hierarchies for object tracking. *Journal of Visual Communication and Image Representation*, **11**, 115–129.

[14] Leeuwen, M.B. and Groen, A. (2005) Motion estimation and interpretation with mobile vision systems. http://www.science.uva.nl/research/ias/research/perception/vehicmotion/.

2

Registration for Video Coding

2.1 Introduction

Digital video is a representation of a natural or real video scene sampled spatially and temporally. The spatial characteristics (shape and number of objects, texture and colour variation, etc.) and the temporal characteristics (object motion, movement of the camera, lighting changes, etc.) are fundamental concepts in video processing. Video coding represents the process of compacting and compressing a video sequence into a smaller number of bits. In order to achieve this compression redundant data are removed, and these data can be features or components that are not useful for the reproduction of the video. Video coding techniques exploit both spatial and temporal redundancies. In the spatial domain there is high correlation between pixels that are close to each other, while in the temporal domain adjacent frames are highly correlated especially when the video frame rate is high. Registration techniques are used during the video coding with the aim to reduce temporal redundancy by exploiting the similarities between neighbouring video frames. Usually, this is achieved by constructing a prediction of the current video frame using motion estimation techniques. Motion estimation is a registration process that compares a region of the current frame with neighbouring regions of the previous reconstructed frame. The aim of this step in the video coding is to find the 'best match' of the neighbouring blocks in the reference frame that gives the smallest residual (obtained by subtracting the prediction from the current frame). Hence, the output of the temporal model is a residual frame and a set of motion vectors describing how the motion was compensated. The motion vectors are then compressed by an entropy encoder in order to remove statistical temporal redundancy in the data.

In this chapter, an overview of registration techniques based on motion estimation methods is presented focusing on applications in video coding. Also different methodologies for understanding the effectiveness of motion estimation techniques are discussed. Most of these evaluation criteria are based on measuring the error

Image, Video & 3D Data Registration: Medical, Satellite & Video Processing Applications with Quality Metrics,
First Edition. Vasileios Argyriou, Jesus Martinez Del Rincon, Barbara Villarini and Alexis Roche.
© 2015 John Wiley & Sons, Ltd. Published 2015 by John Wiley & Sons, Ltd.

between the actual and the predicted frames. In this chapter, techniques inspired by the human visual system (HVS) and the perceptual distortion are also presented, while the concepts of quality of experience (QoE) and quality of service (QoS) are discussed in order to analyse how image and video quality is influenced by registration techniques.

2.2   Motion Estimation Technique

Motion estimation for video coding has been extensively researched over the last few decades, and several algorithms have been presented in the literature. The most popular techniques for motion measurement are the 'matching methods'. Based on these techniques, a region of the current frame is compared with the neighbouring area of the previous frame in order to estimate the motion among different frames. The idea is to divide the current frame into macroblocks and to find, for each block, the one in the previous frame that best matches the current block based on distance criteria. The location of the best match indicates the motion vector, which represents the displacement of a macroblock between two frames. Consequently, the pixels belonging to the same macroblock are subject to the same displacement, and the whole moving area is described by one single vector.

These techniques can be further separated depending on the domain they operate. Matching methods can operate either in the spatial or in the frequency domain. In this chapter, we focus on the first case and the state-of-the-art 'block matching' algorithms used in the most video coding standards (e.g. MPEG4, H.264 and H.265) are analysed.

Not all the techniques used to perform the motion estimation are block based. The 'non-matching' techniques were the first motion estimation algorithms, and they try to exploit the relationship between temporal and spatial intensity changes in a video sequence, usually they use features with a statistical function applied over a local or global area in a frame. These methods are also known as indirect methods, such as corner detection that extract certain features and infer the contents of an image. These approaches are further analysed in the following chapters.

2.2.1   Block-Based Motion Estimation Techniques

Block matching using fixed-size blocks has been arguably the most popular motion estimation technique for more than two decades. It is an important component in several video coding standards, including MPEG-1 and MPEG-2, MPEG-4, H.261, H.263, H.264 and H.265, in which 16×16 macroblocks are commonly used.

In block matching techniques, both the reference and the current frame are split into non-overlapping blocks. These algorithms assume that within each block, motion is equal. One motion vector is assigned to all pixels within a block, thus all the pixels perform the same motion and the algorithm searches the neighbouring area of the reference frame. A common design issue is determining the optimum block size. Although the block size is large, small objects will be bunched together and the assumption of

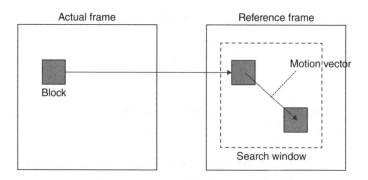

Figure 2.1 Block matching algorithm

uniform motion in a block is invalidated. On the other hand, if the block size is too small, the possibility to determine the best match in an unsuitable position (not the actual motion) is high. In the popular video coding standards, motion estimation and compensation are carried out on blocks of 8×8 or 16×16 pixels.

Block matching algorithms are based on the matching of blocks between two images, the aim being to minimize a disparity measure. Each block in the actual frame is compared to a possible match over a local search window in the reference image (see Figure 2.1).

Video encoder carries out this process for each block in the current frame involving the following steps:

1. Calculate the energy of the difference between the current block and a set of search positions in the previous frame.
2. Select the best match, which is the position that gives the lowest error.
3. Calculate the difference of the current block and the best match region. This step is known as *Motion Compensation*.
4. Encode and transmit (or store) the difference block (error).
5. Encode and transmit (or store) the motion vector that indicates the position of the matching region relative to the current block position.

In order to reconstruct the predicted frame, the video decoder performs the following steps.

1. Decode the difference block and the motion vector.
2. Select the region that the motion vector indicates in the previous frame and add the difference block.

The matching of one block with another is based on the output of a cost function. The block that results in the least cost is the one that matches the closest to current block.

There are different cost functions, and one of the most popular is the sum of absolute error (SAE),

$$\text{SAE} = \sum_{i=0}^{N-1} \sum_{j=0}^{N-1} \left| C_{ij} - R_{ij} \right|$$

where N is the block side, C_{ij} is the ij sample of the current block and R_{ij} is the ij sample of the reference block, where R_{00} and C_{00} are the top-left sample in the current and reference areas, respectively.

Another commonly used matching criterion for the block-based motion estimation is the mean absolute error (MAE), which provides a good approximation of residual energy.

$$\text{MAE} = \frac{1}{N^2} \sum_{i=0}^{N-1} \sum_{j=0}^{N-1} \left| C_{ij} - R_{ij} \right|$$

The mean square error (MSE) is also used providing a measure of the energy remaining in the difference block. MSE can be calculated as follows:

$$\text{MSE} = \frac{1}{N^2} \sum_{i=0}^{N-1} \sum_{j=0}^{N-1} (C_{ij} - R_{ij})^2$$

2.2.1.1 Full-Search

In order to find the best match in the reference frame, the Full-Search algorithm performs a comparison (e.g. computing SAE) of the current block with every possible region of the reference frame. In practice, a good match can be found in the neighbourhood of the block position; thus, the search is limited to a search window centred on the current block position.

Full motion estimation computes the selected disparity measure at each possible location in the search window and it is computationally intensive, especially for large search windows. The algorithm can use a raster (Figure 2.2(a)) or a spiral order (Figure 2.2(b)), starting from the centre position, to scan all the possible locations. The spiral scan method has the advantage that a good match has a higher possibility to occur near the centre of the search region. In this case, a threshold could be used to terminate the comparison process, reducing the required processing power and time. Generally, the size of a search window is 31×31 pixels, and it stands for the most common used dimensions by the block matching techniques [1].

2.2.1.2 Fast-Search: Reduction in Search Position

The aim of fast-search motion estimation algorithms is to reduce the complexity and to provide a more efficient method to determine the best match. In fact, the full-search

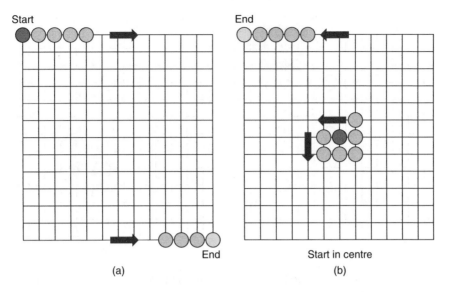

Figure 2.2 Full-Search algorithm using a (a) raster and a (b) spiral order

method is too computationally expensive and it is not suitable for real-time applications. The main disadvantage of fast algorithms is the high probability to find a local minimum instead of the best match that gives the true minimum residual. To the contrary, the Full-Search algorithm is guaranteed to find the global minimum of the disparity measure. As a result, the compression ratio of the difference block is higher for the Full-Search algorithm compared to the fast methods. Since 1981, numerous alternative fast algorithms for block matching motion estimation have been proposed, of which the most popular algorithms are presented in the following sections.

Three-Step Search
This fast algorithm was not developed necessarily for three steps; it can be carried out with different number of steps depending on the search window [2]. The relationship between size and steps is given by size $= 2N{-}1$ (N-step search).

The steps are as follows:

1. Search for the location $(0,0)$ and perform the comparison operation.
2. Determine the step size $S = 2N{-}1$.
3. Search and calculate the disparity measure for all the eight locations $\pm S$ pixels around location $(0,0)$.
4. Select the location with the smallest comparison criterion (SAE) and set it the new search origin.
5. Set $S = S/2$.
6. Repeat steps 3–5 until $S = 1$.

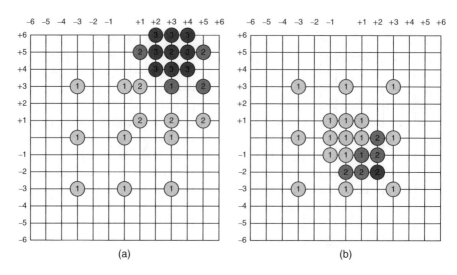

Figure 2.3   (a) The Three-Step Search (TSS) and (b) the New Three-Step Search (NTSS)

Figure 2.3 illustrates the algorithm with a maximum displacement of ±6 pixels in both directions with an initial step size of 3. After each stage the step size is decremented, allowing the algorithm to halt in three steps (hence its name).

The Three-Step Search (TSS) algorithm uses a uniformly allocated checking point pattern in its first step, which becomes inefficient for the estimation of small motions. Several modifications of the original TSS scheme have been proposed. A New Three-Step Search (NTSS) algorithm is proposed [3], and it employs a centre-biased checking point pattern in the first step, which is derived by making the search adaptive to the motion vector distribution. Additionally, the algorithm uses a halfway-stop technique in order to fast identify and estimate the motion of stationary and quasi-stationary blocks. In Figure 2.3(b), an example of search path of NTSS algorithm is shown. If the minimum measured value of the block difference occurs in the first step, at the centre of the search window, the search stops. If the point with the minimum block difference in the first step is one of the eight search points surrounding the window centre, the search in the second step will be performed only at the eight neighbouring points of the minimum and then it stops. Otherwise, the search goes as for TSS.

2D Logarithmic Search

In 1981, the concept of block-based motion estimation based on 2D Logarithmic (TDL) Search was introduced [4]. The steps of the logarithmic method are listed as follows:

1. Search for the location $(0, 0)$ and perform the comparison operation.
2. Select the initial step size.

3. Search and calculate the disparity measure for all the four positions in the horizontal and vertical directions, S pixels away from the $(0, 0)$ location.
4. Select the best match among the five locations and set it the new search origin.
5. If the best match is the origin (the centre) set $S = S/2$, otherwise S is unchanged.
6. Repeat until $S = 1$ (go to step 3). If $S = 1$ go to the next step (step 7).
7. Search and calculate the disparity measure for all the eight locations immediately surrounding the best match ($S = 1$). The location that minimizes the criterion is the result of the algorithm.

An example of this method is shown in Figure 2.4, all blocks outside the search area are deemed to have distortion function values of $+$infinity, and thus are never chosen. Points $[0, +4]$, $[+4, +4]$, $[+6, +4]$ are the minima at each stage, and $[+7, +4]$ is chosen as the matching block. Numbers indicate the stage during which a candidate block is first evaluated.

One-at-a-Time Search
The One-at-a-Time Search (OTS) algorithm is quite simple, and it performs search first in the horizontal direction using three adjacent search points. The steps are listed as follows:

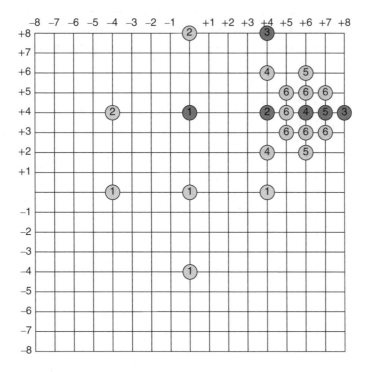

Figure 2.4   The 2D Logarithmic (TDL) Search converging on a position of minimum distortion

1. Set the horizontal origin to $(0, 0)$.
2. Search and calculate the disparity measure for the origin and the two horizontal neighbours.
3. If the middle point has the smallest value, then repeat the above procedure for the vertical direction. Else go to step 4.
4. Select the point with the smallest value and set it the origin. Calculate the disparity measure for the new location and go back to step 3.

OTS is a very simple and effective algorithm with a horizontal stage and a vertical stage. During the horizontal stage, the point of minimum distortion on the horizontal axis is found. Then, starting with this point, the minimum distortion in the vertical direction is found.

Figure 2.5 illustrates the OTS positions of minimum distortion. In Figure 2.5(a), the position of minimum distortion is at $[+2, +6]$, while in Figure 2.5(b) is at $[-3, +1]$.

Cross Search Algorithm

Cross Search Algorithm (CSA) follows the same steps as the TSS method [5]. The main difference is the number of points that are compared. In this case, five points are selected instead of eight. The steps are as follows:

1. Search for the location $(0, 0)$ and perform the comparison operation.
2. Select the four locations at the corners of a rectangular (the edges of a saltire cross 'X') with size S where $S = 2N - 1$. Calculate the disparity measure for the five points and move to the next step.

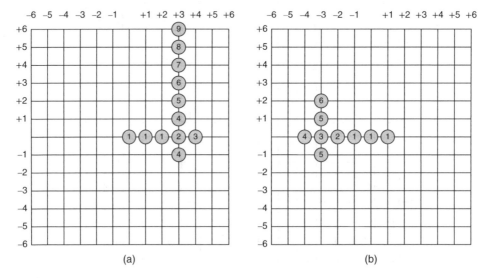

(a)   (b)

Figure 2.5 (a) Match is found at position $(+2, +6)$ and (b) match is found at position $(-3, +1)$

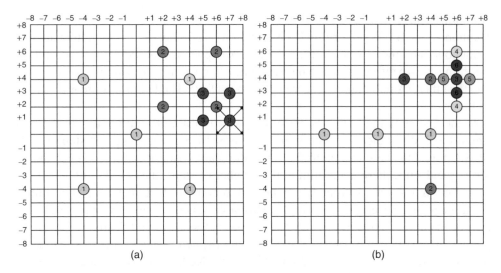

Figure 2.6 (a). Example of the Cross Search Algorithm, arrows illustrate the different patterns used in the final stage. (b) The Orthogonal Search Algorithm (OSA) converging on a position of minimum distortion at (+6, +4)

3. Select the location that minimizes the comparison criterion and set it at the new origin.
4. If $S > 1$, then $S = S/2$ and go back to the step 2, else move to the next stage 5.
5. If the best match is at the top-left or bottom-right location, calculate the selected criterion for the four points at the corners of the rectangle at a distance ± 1. In the case of the best match being at the top right or at the bottom left, calculate the selected criterion for the four points in the immediately horizontal and vertical directions (Figure 2.6(a)).

Orthogonal Search Algorithm
The Orthogonal Search Algorithm (OSA) is a hybrid of the TDL Search and TSS [6]. The OSA has both a vertical stage and a horizontal stage, as illustrated in Figure 2.6(b). After the step size has been initialized to $(d + 1)/2$, where d is the maximum displacement, the centre block and two candidate blocks on either side of the x-axis are compared to the target block. The position of minimum distortion becomes the centre of the vertical stage. During the vertical stage, two points above and below the new centre are examined, and the values of the distortion function at these three positions are compared. The position with the minimum distortion becomes the centre of the next stage. After completing a horizontal iteration and a vertical iteration, the step size is halved if it is greater than 1, and the algorithm proceeds with another horizontal and vertical iteration. Otherwise, it halts and declares one of the positions of the vertical stage to be the best match to the target block.

Diamond Search

Diamond Search (DS) algorithm [7] is similar to NTSS but the search point pattern is a diamond, and there is no limit on the number of steps that the algorithm can take. DS employs two different search patterns, one is large diamond search pattern (LDSP) and the other is small diamond search pattern (SDSP). The DS procedure is described in Figure 2.7.

In the first step, the search is performed with LDSP until the point with the minimum block difference is found at the centre of LDSP. Then, the last step uses the SDSP around the new search origin. Since the search pattern is not too small and not too big and since there is no limit to the number of steps, the DS algorithm can find global minimum very accurately. DS proved to be best block matching algorithm of the last century. Several more complex DS-based schemes have been proposed as an improvement on the result of DS. For example, Cross Diamond Search (CDS) incorporates the cross search pattern at the first step [8], while the Small-Cross-Diamond Search (SCDS) and Kite-Cross-Diamond Search (KCDS) utilize the small cross and kite search patterns combined with a halfway stop technique in order to reduce the computational cost for video sequences with low motion activity, such as videoconferencing [9, 10].

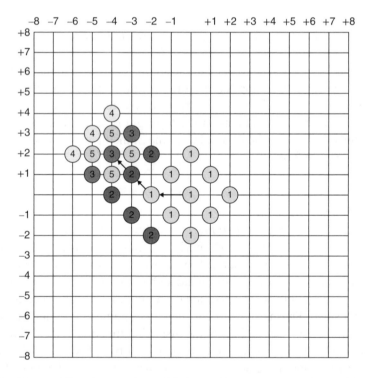

Figure 2.7 Diamond Search procedure. This figure shows the Large Diamond Search Pattern and the Small Diamond Search Pattern. It also shows an example path to motion vector $(-4, 2)$ in five search steps: four times of LDSP and one time of SDSP

Adaptive Rood Pattern Search

In order to achieve high speed and accuracy, it is useful to consider different search patterns, according to the estimated motion behaviour. In fact, we can observe that in case of small motions a small search pattern, made up by compactly spaced search points, is more suitable than a large search pattern as only few positions around the search window have to be checked. In the other case, when the motion is big, the small pattern is not a good choice as it tends to be trapped into local minima along the search path, leading to wrong estimations. Adaptive Rood Pattern Search (ARPS) algorithm assumes that the macroblocks adjacent to the current one belong to the same moving object, and they have similar motion vectors. Hence, the motion of the current macroblock can be predicted by referring to the motion of the adjacent blocks in the spatial and/or temporal domains.

The ARPS consists of two sequential search stages: the first stage is the initial search; the second stage is the refined local search. In the first stage the pattern size is dynamically determined for each macroblock based on the predicted motion behaviour using an adaptive rood pattern (ARP) with adjustable rood arm. The shape of the rood pattern is symmetrical, with four search points located at the four vertices, as shown in Figure 2.8. The step size is the distance between any vertex point and the centre. A cross shape has been chosen based on the observation that the distribution of the motion vector is higher in horizontal and vertical directions as most of the

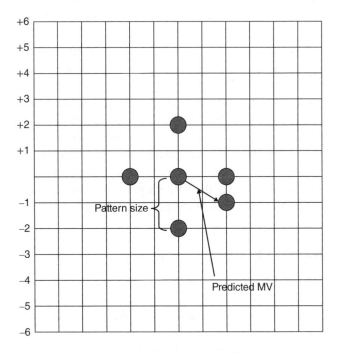

Figure 2.8   Adaptative rood pattern (ARP)

camera movements are in these directions. In addition to the four-armed rood pattern, it is desirable to add the predicted motion vector into ARP because it is very likely to be similar to the target motion vector. In this way, we are going to increase the probability to detect the vector in the initial stage. In the refined local search stage, a unit-sized rood pattern (URP) is exploited repeatedly, until the final motion vector is found.

The ARPS algorithm is summarized in the following steps [11]:

1. The motion vector for the current block is predicted by referring to the left adjacent block. The length of the arm Γ of the ARP is determined as follows:

```
if the current block is a leftmost boundary block ,
      Γ = 2;
Else
      Γ = Max{|MV_predicted(x)|, |MV_predicted(y)|}.
```

2. The centre of the ARP is aligned with the centre of the search window and the four search points plus the position of the predicted motion vector are checked. The length of the four arms is equal to Γ.
3. Set the centre point of the URP at the minimum matching error point found in the previous step and check its new points. If the new minimum matching error point is on any vertex point repeat this step, otherwise the motion vector is found and it corresponds to the minimum matching point identified in this step.

In case of small motion contents, a further improvement in the algorithm can be obtained using the Zero Motion Prejudgment. The matching errors between the current block and the block at the same location in the reference frame is computed and compared with a predefined threshold T. If the matching error is smaller than T, then the current block will be a static block and the research will stop [11].

Based on ARPS algorithm, a new fast and efficient motion estimation algorithm called Enhanced Adaptive Rood Pattern Search algorithm has been proposed. It uses spatial correlation to predict the initial search centre and incorporates an early termination [12]. The main advantage of the ARPS algorithm is its simplicity and regularity. These characteristics are very attractive especially for hardware integrations.

2.2.1.3 Variable-Size Block-Based Techniques

Motion estimation for video coding differs from general motion estimation in several aspects. First, for video coding, the performance of a motion estimator can be assessed using a distortion measure, whereas for the general motion estimation problem, such measures may not be meaningful. Second, the displacement vector field (DVF) for video coding has to be transmitted; therefore, the best performance in terms of distortion using only the available number of bits to encode the vector field is required to be achieved by the coder. Third, for video coding, the resulting vector field does not necessarily correspond to the real motion in the scene. In contrast, the goal of the general motion estimation problem is to find the 2D projection onto the image plane of the real-world 3D motion [13].

Most of the current video coding standards are based on a fixed block-size motion estimation and compensation. The main advantage is the simplicity because they do not need to encode the position and the size of the blocks. There are also standards (H.263 and H.264) that allow splitting of the 16×16 blocks into 8×8 blocks (or 4×4) if that is going to reduce the transmitted error significantly due to the fact that one motion vector is sufficient for regions with small motion, whereas in regions with complex motion the vector field should be denser. As a result, a compact representation of the vector field can be achieved. A lot of work has been done in this area, and we give a description of the fundamental variable block matching algorithms below.

Quad-Tree Techniques
Before we describe some state-of-the-art attempts to find the DVF using variable block size, we would like to point out that, generally, it is suggested for a quad-tree representation of a frame's contents to first apply fixed block size segmentation and then to use a quad-tree structure of three or four levels [14]. This is preferable when motion estimation is used mainly for video coding, but in case of real-motion measurement this becomes a disadvantage.

Let us consider a generalized block-matching motion estimation technique, which combines spatial nonlinear transformations with quad-tree segmentation [15]. This method segments a frame into variable-size blocks and estimates for each one a translational vector and a set of transformation parameters that define the mapping that minimizes the mean-square error between a block in the current frame and an irregular quadrilateral in the previous frame.

A wide variety of algorithms for efficient quad-tree coding can be found in the literature [13, 16–18]. In this case, a Lagrange multiplier method is used to construct quad-tree data structures that optimally allocate rate with no monotonic restrictions. The suggested technique uses a stepwise search to find the overall optimal quad-tree structure. The basic idea is that there is no need to exhaustively consider all possible rate allocations in order to identify the optimal quad-tree in the set of all possible quad-trees. An efficient coding algorithm must minimize both rate and distortion. Using a Lagrange multiplier $\lambda \geq 0$, points in the rate-distortion plane can be found by minimizing the objective function:

$$\min\{d_k + \lambda b_k\}$$

where b_k is the available number of bits for region k and d_k is the resulting distortion. This is also known as the principle of separable minimization.

A quad-tree structure has been used to apply variable-size block-matching motion estimation and compensation for object-based video coding [19]. An example of this

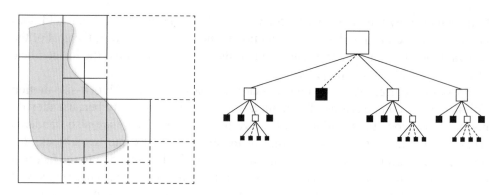

Figure 2.9   Quad-tree structure example

technique is shown in Figure 2.9: the shape and the motion of the objects are described by a quad-tree structure. Some macroblocks are constructed over the shape description and then further extended by one macroblock to cater for movement of the objects between frames.

The block matching operation is performed using small square blocks; in this example 4 × 4 blocks are used. Displacement information is stored as simple bit vectors for those positions which, when the blocks are matched, correspond to a MAE less than a predefined threshold. Blocks are then merged in a quad-tree manner, and the block number depends on the initial threshold. What we obtain is a tree that includes both solid and dotted lines. It is possible to encode efficiently this structure using 1 bit per node, '0' for the leaf and '1' for the internal node, obtaining a tree description using 28 bits. We can minimize the description cropping the blocks that are outside the object mask.

Summarizing, variable block-size-based decomposition of a frame is a good compromise between the complex object-oriented approaches and the simple fixed block-size-based approaches. In addition, when a quad-tree structure is used, the overhead is much lower than the bit rate required to transmit the fine segmentation mask of an object-oriented scheme. All the above techniques operate in the spatial (pixel) domain, and exhaustive or fast block-matching motion estimation algorithms are utilized to obtain the motion field. Furthermore, the MSE or other similar metrics are used as the criterion to split a block into four sub-blocks according to the quad-tree segmentation. The quad-tree structure is an efficient way of segmenting a frame into blocks of different sizes. However, the motion estimation algorithms and the splitting criteria that are mainly used make the techniques able to achieve high compression ratios and consequently very low bit rates.

Hierarchical Techniques
There are a number of complex motion estimation techniques that have the potential to offer performance improvements over conventional block matching techniques. They

are based on the same principles, but they take a more general view of the motion measurement problem. The multiresolution structure, also known as pyramid structure, is very powerful computational configuration for image processing tasks. In fact, it has been regarded as one of the most efficient block matching methods and is mostly adopted in applications that require high-resolution frames or large search areas.

Generally, hierarchical algorithms search a sub-sampled version of a frame, followed by successively higher resolutions. Hierarchical block matching techniques attempt to combine the advantages of large blocks with those of small blocks. The reliability of motion vectors is influenced by block size; and the larger the blocks, the more likely to track actual motion and the less likely to converge on local minima. Although such motion vectors are reliable, the quality of matches for large blocks is not as good as that of small blocks.

Hierarchical block matching algorithms exploit the motion estimation capabilities of large blocks and use their motion vectors as a starting point for searches in smaller blocks [20]. Initially, large blocks are matched, and the resulting motion vector provides a starting point for the smaller blocks. In more details, a hierarchical block matching algorithm is started by matching a large target block with a similar-sized block in the past frame. The resulting motion vector is then used as an estimate for smaller targets and candidate blocks. At each stage, the obtained estimate was used as a starting point for the next stage, as is illustrated in Figure 2.10(a). The size of the block was reduced and a match was found at each stage of the algorithm, until the matching blocks were of the desired size. In Figure 2.10(b), a three-level hierarchical motion estimation algorithm is showed. Levels are numbered from 2 to 0, where 2 represents the coarsest level and 0 represents the finest level. At levels 2 and 1, the image is sampled with a ratio 16:1 and 4:1, respectively. A TSS algorithm can be used to find an initial estimation for the next level. Additionally, a low-pass filtering by averaging,

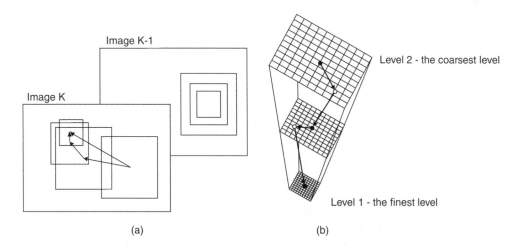

Figure 2.10   (a, b) A tree-level hierarchical search

for example, the pixels at the first level is applied to reduce the noise influence [21]. Usually, a two- or three-level hierarchical search is adopted and in order to improve the performance, multiple search schemes at each step can be applied resulting in many modifications of the original approach [22–25].

2.3   Registration and Standards for Video Coding

Many international video compression standards use motion estimation techniques to remove the temporal redundancy between the current frame and the reference frame. Motion estimation provides a coding mechanism that results high compression ratio. An efficient implementation of the motion estimation algorithms is very important because it represents the most computationally intensive part in a video encoder (50–90% of the entire system). The block matching algorithms, presented in the earlier sections, have been widely adopted and selected as the motion estimation module due to their effectiveness, performance and implementation simplicity. In the following two sections, we will discuss the main features of the latest video coders, highlighting the techniques that are used to perform the motion estimation task.

2.3.1   H.264

One of the most recent and most popular international video coding standard of the ITU-T Video Coding Experts Group and the ISO/IEC Moving Picture Experts Group [26] is H.264/MPEG-4 AVC. It represents a video compression technology, and it is suitable for the full range of video applications including low bit rate wireless video applications, standard-definition and high-definition broadcast television, and video streaming over the Internet.

The video is compressed using a hybrid of motion compensation and transformation coding. Video coding algorithms are used in order to remove different types of redundancy inherent in video:

- Spatial redundancy.
- Temporal redundancy.
- Perceptual redundancy.
- Statistical redundancy.

In Figure 2.11, the algorithm used to reduce all these redundancies is visualized, highlighting the components that use registration techniques (block matching and motion estimation) to reduce mainly the temporal redundancy.

The decoding process was standardized in the video coding standard, but not the encoding process. This approach is very flexible and lets to use different technologies in the encoder side that is more complex than the decoder. Similar to other hybrid video coding standards, H.264 is a block-based video coding standard, where the

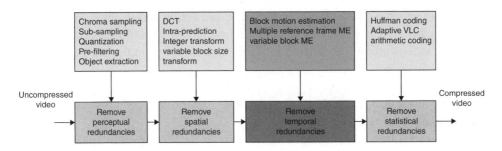

Figure 2.11   Component for the compression process in H.264

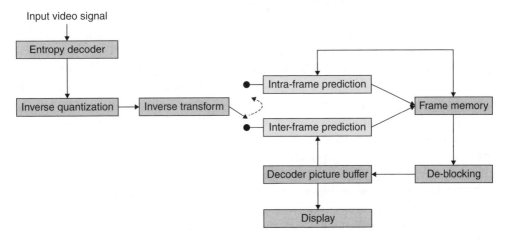

Figure 2.12   Components in H.264 decoder

video is encoded and decoded in macroblocks per time. As in MPEG-2, the intra (I) frame is coded without using other frames for prediction, predictive (P) frame is coded using previously coded frames for prediction and bi-predictive (B) frame uses two previously coded frames for prediction. In H.264, other two types of frames are defined: switching-I (SI) and switching-P (SP) frames, used in streaming applications [27].

In Figure 2.12, the key features of the H.264 standard are shown. The intra-prediction block and the de-blocking are new features, not present in MPEG-2. The intra-prediction exploits spatial redundancy better than the previous standards. A macroblock is coded as an intra-macroblock when temporal prediction is not possible, for example, for the first frame in a video or during the scene changes. The prediction is determined using the neighbouring pixels in the same frame due to their similarities.

The de-blocking improves the perceptual video quality and the prediction efficiency in inter-frames. It is applied after a macroblock has been reconstructed to the block edges, but not in the frame or the block boundaries.

The block in which the motion-compensated prediction and the motion estimation algorithms are used is the inter-frame prediction block. This block is present in all the earlier video coding standards. The H.264 extends it by employing the following features:

- Variable block size for motion compensation.
- Multiframe reference for prediction.
- Generalized B frame prediction.
- Use of B frame references.
- Weighted prediction.
- Quarter pixel accuracy for motion vectors.

This advanced video coding does not use the fixed-size block matching algorithms, because it is difficult to accommodate the different changes in object movement within a video frame. H.264 avoids this limitation by the employment of variable block size algorithms and motion vectors with quarter-pixel resolution. For the motion vectors fractional accuracy is used, and this means that the residual error is decreased while the coding efficiency of video coders is increased [28]. It is important to consider that if a frame has a fractional value, the reference block has to be interpolated accordingly. The design of this interpolation filter takes into account many factors, such as complexity, efficiency and visual quality [29].

Regarding the size of the blocks, a macroblock can be coded as one 16×16, two 16×8, two 8×16 or four 8×8 blocks. Each 8×8 block can be coded as one 8×8, two 4×8, two 8×4 or four 4×4 sub-macroblocks (see Figure 2.13).

In the multiple reference motion estimation (MRF-ME), an I-frame can use as many as 16 different reference frames. The actual number or reference frame is

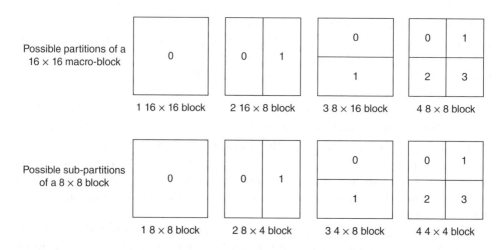

Figure 2.13   Possible partition of macroblocks and possible sub-partitions of a block

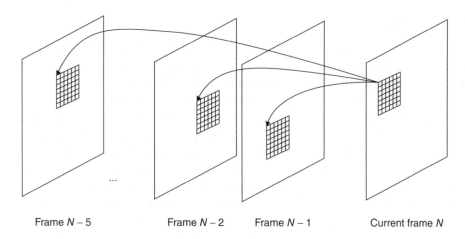

Frame $N-5$ Frame $N-2$ Frame $N-1$ Current frame N

Figure 2.14   Multiple reference motion estimation

only limited by the buffer size in a particular profile and level. And similarly, a motion-compensated bi-prediction macroblock in B slices is predicted from any two predictions, not the one previous or following as in MPEG-2 standard, and also a B frame can be used as reference picture. A scenario of MRF-ME is shown in Figure 2.14.

These features contribute to make the motion compensation tool more efficient. The main reasons for using MRF-ME prediction instead of the single reference prediction are the following [30, 31]:

- *Repetitive motions:* Due to the repetitive nature of the motion, there are better correspondences of the same object/texture in several frames ago.
- *Uncovered background:* A moving object can cover some parts of the frame. As the object moves, the background, which is uncovered, may not find a good match from the previous frame but could find a good match many frames ago when the object did not cover the background.
- *Alternating camera angles:* These angles switch back and forth between two different scenes.
- *Sampling:* If we consider that an object moves with a non-integer pixel displacement, the sampling positions of the object in different frames may not be the same. In this case, the current block may get a better match to a block in previous reference frames.
- *Shadow and lighting changes:* Different shadowing, lighting conditions or reflections can lead an area or a moving object to not have exactly the same pixel values in adjacent frames.
- *Camera shaking:* When a camera is moving up and down, the current frame may better correspond to one frame ago.
- *Noises in the source signal:* These are produced by the camera and other factors.

2.3.2   H.265

The increase in available services, the growing popularity of HD videos and the introduction of formats, such as 4k × 2k or 8k × 4k resolution, lead to a need to improve the coding efficiency that can be obtained by H.264/MPEG-4 AVC.

The most recent joint video project of the ITU-T VCEG and ISO/IEC Moving Picture Experts Group (MPEG) standardization organizations is the High-Efficiency Video Coding (HEVC). In April 2013, the first version of their common text specification [32] was approved as ITU-T Recommendation H.265 and ISO/IEIC 23008-2 (MPEG-H, Part 2). New features extend the standard in order to support several different scenarios, such as 3D, stereo video coding and scalable video coding.

HEVC has been designed to perform all the essential applications of H.264/MPEG-4 AVC but aiming to achieve multiple goals, such as increased video resolution, coding efficiency, data loss resilience, and making easy the integration of a transport system. It also aims to increase the use of parallel processing architectures.

Technically, the HEVC employs the same hybrid block-based approach used in H.264/MPEG-4 AVC. As it can be seen in Figure 2.12, the basic source-coding algorithm is a hybrid of inter-picture prediction, in order to exploit temporal dependencies, intra-picture prediction to exploit the spatial dependencies, and the transform coding of the prediction of residual signals to exploit further spatial dependencies. But in contrast to H.264/MPEG-4 AVC, which supports prediction block size from 4×4 to 16×16 and transform block size mainly of 4×4 and 8×8 samples, the HEVC allows to use a larger range of block sizes for both intra-picture and inter-picture predictions, as well as for transform coding. An important consideration is that there is no unique element that provides most of the improvements comparing with the previous standards, but many small changes lead to a considerable high gain.

An essential innovation from the new standard is that it allows dividing the prediction residual blocks recursively into smaller transform block by using a nested quad-tree structure referred as residual quad-tree (RQT) [33]. Each input picture in the codec is partitioned in a regular grid of disjoint square blocks $N \times N$ samples, referred as coding tree blocks (CTBs). Each of these CTB represents the root of a quad-tree and can be subdivided into smaller square blocks, called coding blocks (CBs). In version 1 of H.265/HEVC, CB sizes from 8×8 to $N \times N$ are supported, where the maximum value for N is 64. For each CB, a prediction mode is signalled as intra or inter, according to the used prediction. Each CB represents a leaf of the quad-tree, and is further subdivided into prediction blocks (PBs) and transform blocks (TBs). The subdivision of a CB into TBs for residual coding is performed by using a second, nested quad-tree with the CB as its root.

The motion vector is predicted by using the advanced motion vector prediction (AMVP) method that includes the derivation of several most probable candidates based on the data coming from the reference picture and adjacent PBs. It is possible to use also a merge mode for motion vector coding through the inheritance of MVs from both temporally and spatially neighbouring PBs.

As in H.264/MPEG-4 AVC, a quarter-sample precision is used for the motion vectors. For the interpolation of fractional sample positions, 7-tap or 8-tap filters are used, in contrast to the 6-tap filter of half-sample positions followed by linear interpolation for quarter-sample position that are used in H.264/MPEG-4 AVC. As the previous standard, a multiple reference picture is considered. For each PB, one or two MVs can be transmitted, resulting, respectively, in unipredicted and bipredicted coding.

2.4 Evaluation Criteria

In order to assess the performance of a given motion estimation algorithm, some quantitative error measurements are required. The key issue on selecting a quantitative error measure relates to whether or not ground truth is available, which unfortunately is not always the case. In this section, three different error measures are presented. One of which is for the case of known true motion fields, and the others for unknown.

2.4.1 Dataset

To illustrate the efficiency and provide a comparative study of the motion estimation algorithms discussed in this chapter some well-known experimental frameworks can be utilized. For example, the broadcast resolution (720 × 576 pixels, 50 fields per second) test sequences 'Mobcal', 'Basketball', 'Garden', 'Foreman' and 'Yos' can be used (Figure 2.15). These sequences are utilized to evaluate MPEG video coding standards, providing a significant variety of different motion contents. 'Mobcal' contains mainly translational object motion, with the exception of the object 'ball' that is rotating, slow camera pan, synthetic textures, occasional occlusions and lacks any significant scene depth. 'Basketball' is a more active scene containing fast object motion, slow camera motion and a fairly complex background. 'Garden' is characterized by a significant scene depth and projective motion, natural and multicoloured textures as well as significant occlusions due to the foreground 'tree' object. 'Foreman' is a head-and-shoulders scene with varying foreground and low background motions. 'Yos' is an artificial sequence characterized by scene depth, projective motion and large displacements without occlusions. In this sequence, ground truth is available and the real motion is known a priori.

One of the latest datasets in video coding focusing on HD and 4K quality (generally 3840 × 2160) is SJTU 4K [34], offering a set 15 of 4k video test sequences, in uncompressed formats: YUV 4:4:4 colour sampling, 10 bits per sample and YUV 4:2:0 colour sampling, and 8 bits per sample formats with frame rate 30. Another dataset that can be used both for video coding and action recognition is the Human Motion DataBase (HDMB) [35], with more than 7000 video clips extracted from a variety of sources ranging from digitized movies to YouTube (see Figure 2.16).

In order to evaluate the performance of quality assessment algorithms, another popular database is the LIVE Video Quality Database [36]. It uses 10 uncompressed

(a)

(b)

(c)

(d)

(e)

Figure 2.15 Three frames of 'Mobcal' sequence (a), 'Basketball' sequence (b), 'Garden' sequence (c), 'Foreman' sequence(d) and 'Yos' sequence (e). *Source*: https://media.xiph.org /video/derf/

Figure 2.16   Sample frames from HDMB

high-quality videos with a wide variety of content as reference videos (Figure 2.17). A set of 150 distorted videos were created from these reference videos using four different distortion types, such as MPEG-2 compression, H.264 compression, simulated transmission of H.264 compressed bitstreams through error-prone IP networks and through error-prone wireless networks. Each video in the LIVE Video Quality Database was assessed by 38 human subjects in a single stimulus study with hidden reference removal, where the subjects scored the video quality on a continuous quality scale. The mean and variance of the difference mean opinion scores (DMOS) obtained from the subjective evaluations, along with the reference and distorted videos, are available as part of the database.

Figure 2.17 Sample frame from LIVE Video Quality Database

2.4.2 *Motion-Compensated Prediction Error (MCPE) in dB*

A more indirect approach to the validation of the estimated motion field is required in case ground truth is not available. In such cases, the displaced frame difference (DFD) is often used [37] to infer the quality of the motion estimation result. The DFD of any point can be expressed as

$$\text{DFD}_f(x, y) = \left| I_f \begin{pmatrix} x \\ y \end{pmatrix} - I_{f+1} \begin{pmatrix} x + u_e \\ y + v_e \end{pmatrix} \right|$$

In order to provide a quantitative measure of the quality of the estimated motion field, the root mean square (RMS) DFD is computed as

$$\text{DFD}_f = \sqrt{\frac{\sum_{i=0}^{N_R-1} \left(\text{DFD}_f(x_i, y_i) \right)^2}{N_R}}$$

Examples of DFDs after motion compensation are shown in Figure 2.18, where the mid-grey value indicates low error while bright and dark areas correspond to positive and negative errors.

Performance can be assessed applying motion compensation using the estimated motion parameters and computing either the DFD or the MSE. The above quantitative measure can also be expressed in decibels (dB) using the peak signal-to-noise ratio (PSNR):

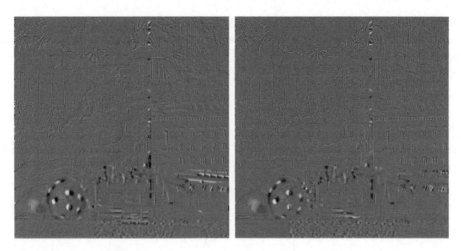

Figure 2.18   An example of motion-compensated frame differences corresponding to 1 dB gain (right over left frame)

$$PSNR = 10\log_{10}\frac{255^2}{\frac{1}{M \times N}\sum_{i=0}^{M-1}\sum_{j=0}^{N-1}(I_f(i,j) - I_{f+1}(i,j))^2}$$

where I_f and I_{f+1} are the intensities of two successive frames. In Figure 2.18, an example of 1 dB gain is shown providing an illustration of the amount of error corresponding to that performance. It should be mentioned that this gain depends on the image quality, and for more detailed frames the same amount of gain is significantly less visible.

2.4.3   Entropy in bpp

The zero-order entropy of the motion-compensated prediction error (MCPE) and the zero-order 2D entropy of the motion vector field are used as an indicator of the compression potential of the resulting motion-compensated prediction error and as an indicator of smoothness, coherence and ultimately compression potential of the motion vector field, respectively.

The entropy equation provides a way to estimate the average minimum number of bits required to encode a string of symbols, based on their frequency of occurrence.

$$H(X) = -\sum_{i=0}^{N-1} p_i\log_2 p_i$$

In the above entropy equation, p_i is the probability of a given symbol. Entropy provides a lower boundary for the compression that can be achieved by the data representation (coding) compression step. Rate-distortion plots (MSE vs. motion vector field entropy in bits/pixel) can also be used to provide a simultaneous representation of the motion-compensated prediction error and the 2D entropy of the motion vector field.

2.4.4   Angular Error in Degrees

In the case that the real motion field is known, the error measure referred as the angular error can be used [38] and is given by

$$
\Psi \left([x \ y]^T \right) = \cos^{-1} \left[\frac{1}{\sqrt{1 + u_c^2 + v_c^2}} \begin{pmatrix} u_c \\ v_c \\ 1 \end{pmatrix}^T \cdot \frac{1}{\sqrt{1 + u_e^2 + v_e^2}} \begin{pmatrix} u_e \\ v_e \\ 1 \end{pmatrix} \right]
$$

where (u_c, v_c) and (u_e, v_e) represent the correct and the estimated motion values, respectively.

To form an estimate for the whole image, the mean value can be calculated as

$$
\Psi = \frac{\sum_{i=0}^{N_R-1} \Psi_i}{N_R}
$$

where N_R indicates the number of pixels for which a motion estimate is available. An example of 12° angular error gain in average is shown in Figure 2.19 where dark areas indicate high error while brighter pixels correspond to lower error.

It should be noted that although there is a clear correlation between the motion field estimation quality and MCPE [38], other factors not directly relating to the quality of the estimates could affect the MCPE measure. Methods that rely extensively on minimizing the MCPE to provide motion measurements without giving much attention

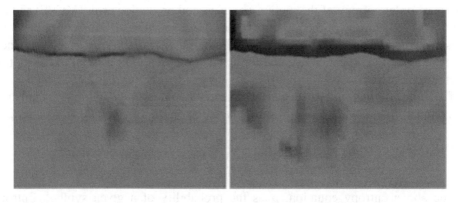

Figure 2.19   An example of 12° angular error gain (right over left frame)

to factors such as spatial smoothness are likely to result in an impressive DFD error with less impressive motion estimates. Therefore, the DFD error should support the results obtained using the angular error measure and provide qualitative assessment of a given algorithm of real-world image sequences.

2.5   Objective Quality Assessment

One of the main targets in the design of modern multimedia systems is to improve the video quality perceived by the user. As shown in the previous section, most of the evaluation criteria for the motion estimation algorithms are based on measuring the error between the actual and the predicted displacements. But these measures do not necessarily correlate well with quality as it is perceived by an end-user [38].

In order to obtain such quality improvement, the availability of an objective quality metric that represents well the human perception is crucial. Objective quality assessment methods based on subjective measurements are using either a perceptual model of the HVS or a combination of relevant parameters tuned with subjective tests [39, 40].

The objective quality metrics can play a variety of roles: they can be used to dynamically monitor and adjust image quality; to optimize algorithms and parameter settings of image processing systems; or to benchmark image processing algorithms. They can be classified according to the availability of an original image, which is compared with the distorted image. Most of the existing approaches are known as full reference, which means that the distortion-free reference image is assumed to be known. In most practical applications, the reference image is only partially available in the form of a set of extracted features, sent to the receiver through an ancillary channel evaluating the quality of the distorted image. This is referred to as reduced-reference (RR) quality assessment. Furthermore, motion information or video temporal model can affect the quality of the video sequence. Considerable resources in the HVS are devoted to motion perception, and it is hence essential for video quality metrics to incorporate some forms of motion modelling. Aiming at improving the evaluation of video quality, the objective quality metrics can be modified and improved by taking into account the motion information of the video sequence introducing the temporal masking effects.

2.5.1   Full-Reference Quality Assessment

In the literature, the most used full-reference objective video and image quality metric is PSNR, which was already defined in Section 4.1 as

$$PSNR = 10\log_{10}\frac{L^2}{MSE}$$

where MSE is the mean-squared error, defined as

$$\text{MSE} = \frac{1}{N} \sum_{i=1}^{N} (x_i - y_i)^2$$

where N is the number of pixels in the image or video signal, and x_i and y_i are the ith pixels in the original and distorted signals, respectively. L is the dynamic range of the pixel values. For instance, for an 8 bits/pixel signal, L is equal to 255. PSNR is widely used because of its clear physical meaning and its simplicity.

However, this metric does not correlate well with subjective quality measures, and hence it has been widely criticized. In the last decades, a big effort has been made to develop a full-reference metric, which takes advantage of known characteristics of the HVS. New paradigms consider that the HVS is highly adapted for extracting structural information.

2.5.1.1   FR Quality Assessment Using Structural Distortion Measures

A new philosophy in designing image and video quality metrics is considered. Natural image signals are highly structured, and spatially approximate pixels have strong dependencies, which keep relevant information about the structure of the object in the scene [41]. A new philosophy has been proposed based on the concept that the HVS extracts structural information from the viewing field [42, 43]. It aims to move from error measurement towards structural distortion measurement. To demonstrate the validity of this assumption, we consider Figure 2.20, where the original image is impaired with different distortions to yield the same MSE. It is interesting to notice that images with identical MSE have different perceptual quality Q.

Wang *et al.* in [39] proposed a structural similarity measure. The SSIM index models any distortion as combination of three factors: luminance information l, contrast information c and structural information s,

$$\text{SSIM}(\mathbf{x}, \mathbf{y}) = [l(\mathbf{x}, \mathbf{y})]^{\alpha} \cdot [c(\mathbf{x}, \mathbf{y})]^{\beta} \cdot [s(\mathbf{x}, \mathbf{y})]^{\gamma}$$

where $\alpha > 0$, $\beta > 0$ and $\gamma > 0$ are used to adjust the relative importance of the three contributions, and $\mathbf{x}$, $\mathbf{y}$ are the original and distorted signals, respectively.

The structural information is independent of the illumination, and then, in order to obtain the structural information, the influence of luminance and contrast are separated. The block diagram in Figure 2.21 shows this process.

Another important issue concerns how to apply this index in the case of video sequences. Wang *et al.* in [44] proposed a video quality system in which they consider the luminance and the motion evaluation, as it is shown in Figure 2.22.

Two adjustments are employed: the first adjustment is based on the observation that dark regions usually do not attract attention, hence should be assigned smaller

Figure 2.20   Comparison of 'Lena' images with different types of distortions: (a) original 'Lena' image, 512×512, 8 bits/pixel; (b) multiplicative speckle noise contaminated image, MSE = 225, $Q = 0.44$; (c) blurred image, MSE = 225, $Q = 0.34$; (d) additive Gaussian noise contaminated image, MSE = 225, $Q = 0.39$; (e) mean shifted image, MSE = 225, $Q = 0.99$; (f) contrast stretched image, MSE = 225, $Q = 0.93$; (g) JPEG compressed image, MSE = 215, $Q = 0.28$; (h) impulsive salt-pepper noise contaminated image, MSE = 225, $Q = 0.65$

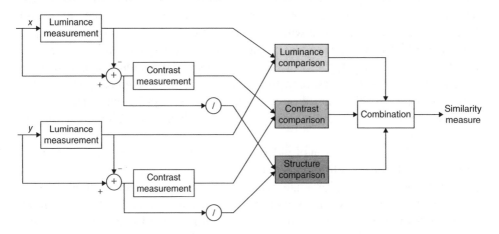

Figure 2.21   Diagram of structural similarity measurement system

weighting values. The second adjustment considers very large global motion, as the distortions are perceived very unpleasant in still or slowly moving videos.

Even if this metric performs very well and shows a good correlation with subjective evaluation, it needs a full-reference image and this is a big limitation for a wide number of applications.

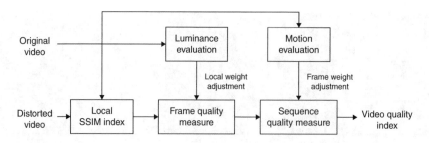

Figure 2.22　Video quality assessment system based on structural information

2.5.2　No-Reference and Reduced-Reference Quality Metrics

For most applications, the reference image is not completely available, it would require a wide amount of storage space and/or transmission bandwidth, and it represents a serious impediment to the practicability of video and image quality assessment metrics. In addition, the original video sequence – prior to compression and transmission – is not usually available at the receiver side, and it is important to rely at the receiver side on an objective video quality metric that does not need reference nor needs minimal reference to the original video sequence. Hence, the full-reference metrics are not always useful, as in many cases, the reference signals are not accessible at the end of transmission system.

The development of automatic algorithms that use less information from a reference signal is an intense focus of research. It is important to evaluate the quality of the received video sequence with minimal reference to the transmitted one [45]. These algorithms can belong to two possible categories: RR and No-Reference (NR) algorithms. In the first category, only some features of the reference signal are accessible; while in the second category, the reference signal is completely inaccessible. These kinds of algorithms are desired especially nowadays because the consumers have become more expert about digital video, and their expectations regarding the QoS have risen significantly.

For closed-loop optimization of video transmission, the video quality measure can be provided as feedback information to a system controller [46]. Figure 2.23 reports a schematic representation of an image/video processing system, consisting of a video encoder and/or a transmission network, with the calculation of a RR quality metric. Reference features are extracted from the original image/video sequence, and these are then compared with the same features extracted from the impaired video to obtain the RR quality metric.

The video transmission network may be lossy, but the RR data channel is assumed to be lossless, in order to guarantee the correct reception of the features at the receiver side.

Instead, the design of No-Reference algorithms is very difficult. This is due to the limited understanding of the HVS and the cognitive aspects of the brain. A number of factors play an important role in a human evaluation of the quality, such as aesthetics,

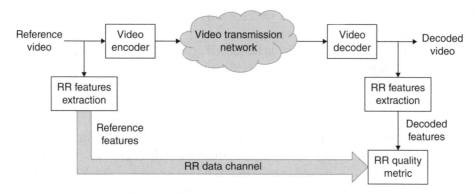

Figure 2.23   Reduced-reference scheme

learning, visual context and cognitive relevance, and each of these elements introduces variability among the observers.

2.5.2.1   RR Algorithms

The human eye is very sensitive to the edge and contour information of an image, that is, the edge and contour information gives a good indication of the structure of an image and it is critical for a human to capture the scene [47].

Some works in the literature consider edge structure information. For instance, in [48], the structural information error between the reference and the distorted image is computed based on the statistics of the spatial position error of the local modulus maxima in the wavelet domain. In [42], a parameter is considered to detect a decrease or loss of spatial information (e.g. blurring). This parameter uses a 13-pixel spatial information filter (SI13) to measure edge impairments. Other metrics consider the Sobel operator [49] for edge detection, as this is one of the most used methodologies to obtain edge information due to its simplicity and efficiency.

A few RR metrics have been proposed, with different characteristics in terms of complexity, of correlation with subjective quality, and of overhead associated with the transmission of side information. The ITS/NTIA (Institute for Telecommunication Sciences/National Telecommunications and Information Administration) has developed a general video quality model (VQM) [50] that was selected by both ANSI and ITU as a video quality assessment standard based on its performance. This general model requires, however, a bit rate of several Mbps (more than 4 Mbps for 30 fps, CIF size video) of quality features for the calculation of the VQM value, which prevents its use as an RR metric in practical systems. The possibility to use spatiotemporal features/regions was considered in [51] in order to provide a trade-off between the correlation with subjective values and the overhead for side information. Later on, a low-rate RR metric based on the full-reference metric [52] ('10 kbits/s VQM') was developed by the same authors. A subjective data set was used to determine the optimal linear combination of the eight video quality parameters in the metric.

The performance of the metric was presented in terms of a scatterplot with respect to subjective data.

The quality index in [45] is based on features that describe the histograms of wavelet coefficients. Two parameters describe the distribution of the wavelet coefficients of the reference image using a generalized Gaussian density (GGD) model, hence only a relatively small number of RR features are needed for the evaluation of image quality.

The RR objective picture quality measurement tool of compressed video in [53] is based on a discriminative analysis of harmonic strength computed from edge-detected pictures, to create harmonics gain and loss information that could be associated with the picture. The results achieved are compared by the authors with a VQEG RR metric [51, 54], and the performance of the proposed metric is shown to be comparable to the latter, with a reduction in overhead with respect to it and a global reduction of overhead with respect to full-reference metrics of 1024:1. The focus is on the detection of blocking and blurring artefacts. This metric considers edge detection that is performed over the whole image, and edge information is not used as side information, but just as a step for further processing the frame for the extraction of different side information.

The quality criterion presented in [55] relies on the extraction, from an image represented in a perceptual space, of visual features that can be compared with those used by the HVS (perceptual colour space, CSF, psychophysical subband decomposition, masking effect modelling). Then, a similarity metric computes the objective quality score of a distorted image by comparing the features extracted from this image to features extracted from its reference image. The size of the side information is flexible. The main drawback of this metric is its complexity, as the HVS model (which is an essential part of the proposed image quality criterion) requires a high-computation complexity.

In [56], an RR objective perceptual image quality metric for use in wireless imaging is proposed. Specifically, the normalized hybrid image quality metric (NHIQM) and a perceptual relevance weighted L_p-norm are designed, based on the observation that the HVS is trained to extract structural information from the viewing area. Image features are identified and measured based on the extent by which individual artefacts are present in a given image. The overall quality measure is then computed as a weighted sum of the features.

The metric in [57] is based on a divisive normalization image representation. No assumptions are made about the type of impairment. This metric requires training: before applying the proposed algorithm for image quality assessment, five parameters need to be learned from the data.

2.5.3   Temporal Masking in Video Quality Assessment

The visual system can be modelled by different steps of signal processing. The first step is the signal processing in the retina. As a consequence of the neural combinations in the retina, luminance signals are bandpass filtered. A contrast sensitivity function (CSF) is defined as the transfer function of the corresponding bandpass filter. A CSF for non-moving image structures is given in [58] as

$$\text{CSF}(f) = (0.2 + 0.45f)e^{-0.18f}$$

where $f = \sqrt{f_x^2 + f_y^2}$ is the absolute value of the spatial frequency.

In the visual cortex, there are further neural combinations of the signals. They can be modelled as a multichannel system that processes the image in subbands of different centre orientations and frequencies. Signals can mask each other in each subband; hence, a stronger signal can make another signal of similar frequency and orientation invisible or at least decrease its visibility. This effect is referred to as spatial masking.

The temporal masking is an important feature of HVS. It would accept higher distortion due to larger temporal masking effects. It is possible to distinguish between two types of changes in a video sequence. The first kind is rapid change, which is used to appear after scene cuts. The consequence is that a significant masking only is visible in the first one or two frames after the scene cut. The second type of temporal masking is more important for the image quality analysis, since it is not limited to single images: it is a reduction of the visibility of coding artefacts due to motion. In psychophysiological studies, moving sinusoidal patterns have been shown to viewers to analyse their visibility depending on the velocity.

The result is that the CSF peak moves towards lower frequencies with higher velocity of the data set test. According to this outcome, higher frequency components of the signal become less visible. It could be modelled in a multichannel manner: in fact, velocity-dependent weighting factors corresponding to samples of the CSF to the different subbands are applied. The temporal masking effects depend on the retinal image and that is considered as a problem. Thus, in order to take advantage of these effects, the velocity would have to be measured relative to the eye movement. If the viewer's eye is able to track a certain motion in a video, it will experience no masking effects at all. Therefore, only an approximation on a statistical basis is possible.

2.5.3.1 Temporal Masking in Video Quality Assessment

The amount of motion in a video scene is measured to judge the effects of temporal masking. To calculate this quantity, the mean of the absolute value μ of motion vectors is used. However, the mean of the absolute value is probably not the best choice, because a high average velocity can correspond to a very homogeneous motion vector field. A homogeneous motion will very likely be tracked by the eye of a viewer, for instance, in the case of a camera pan; hence, it means a very low amount of temporal masking. A better choice could be the standard deviation, where the standard deviation σ of the motion vectors is defined as

$$\sigma = \sqrt{\sigma_x^2 + \sigma_y^2}$$

and where σ_x and σ_y are the standard deviations of the horizontal and vertical components of the motion vectors. σ can be used to measure the scattering of the motion vectors. The meaning is that if σ is large, the motion vector field is inhomogeneous.

In order to determine the temporal quality, motion information is extracted from the reference video sequence in the form of optical flow fields. A block-based motion estimation algorithm is used to evaluate the motion with respect to its previous frame. More details about this algorithm are in Section 1.1.

Temporal masking is an important feature of HVS. Rapid change of the video content between adjacent frames is not perceivable to the HVS. When the background of the video is moving very fast, some image distortions are perceived differently. For instance, severe blurring is usually perceived as very unpleasant kind of distortion in slowly moving video or still images. However, since large perceptual motion blur occurs at the same time, the same amount of blur is not as important in a large motion frame. Such kind of differences cannot be captured by the structural index such as SSIM because they do not consider any motion information.

As the HVS can tolerate better, the distortions in fast-moving regions, spatial distortion, can be measured according to temporal activities of the video, which are calculated as the mean value of the motion vectors in the frame. The temporal activity of the nth frame A_n is defined as

$$A_n = \overline{|mv_{x,n}(x,y)|} + \overline{|mv_{y,n}(x,y)|} \quad 1 \le x \le W, \ 1 \le y \le H \quad (2.1)$$

where $mv_{x,n}(x,y)$ and $mv_{y,n}(x,y)$ are the horizontal and the vertical components of the MV at (x,y) in the nth frame, respectively.

In [59], Yang *et al.* propose a formula to weigh the spatial distortion in order to consider the temporal activity of a video sequence. The weighted index of similarity is

$$I'_s = \frac{I_s}{\left(\beta + \frac{\max(An,\gamma)}{\delta}\right)} \quad (2.2)$$

where I_s is the index of similarity, without taking in account the temporal activity, of the nth frame, and β, λ, and δ are constants determined by experiments.

The video quality metric is obtained by averaging the weighted index of the images over the entire video sequence, computed using (2.2). Some improvements can be obtained selecting salient points in a frame of a video sequence. In fact, under the assumption that the HVS often pays more attention to moving objects than to stationary backgrounds, some salient points can be selected in a frame [60].

2.6 Conclusion

In this chapter registration techniques for video coding applications are presented. The aim of video coding applications is to reduce temporal redundancy in a video sequence by exploiting the similarities between neighbouring video frames. Usually, this is achieved using the motion estimation techniques by reconstructing a prediction of the current video frame.

The first part of this chapter was focused on an overview of the most important motion estimation methods, presenting both the traditional block-based methods and the latest approaches such as the ARPS or the quad-tree techniques. Also, an overview of the latest video coding standards was presented, highlighting the role and the importance of the motion estimation methods and how they are employed. Finally, in the last part of this chapter, different criteria on the assessment of the effectiveness of the motion estimation are presented, also discussing new methods inspired to the HVS.

2.7 Exercises

Implement TSS algorithm in MATLAB in order to calculate the motion vectors of an image when the reference one is known. Compute for the same image the motion vectors using DS method and analyse the results in terms of PSNR. Compare the algorithms in terms of the average number of points searched per macroblock.

References

[1] Ghanbari, M. (1999) *Video Coding*, The Institution of EE, London.
[2] Koga, J., Iinuma, K., Hirano, A. *et al.* (1981) *Motion Compensated Interframe Coding for Video Conferencing*. Proceedings of the National Telecommunications Conference, pp. G5.3.1–G5.3.3.
[3] Li, R., Zeng, B. and Liou, M.L. (1994) A new three-step search algorithm for block motion estimation. *IEEE Transactions on Circuits and Systems for Video Technology*, **4** (4), 438–442.
[4] Jain, J. and Jain, A. (1981) Displacement measurement and its application in interframe image coding. *IEEE Transactions on Communications*, **29** (12), 1799–1808.
[5] Ghanbari, M. (1990) The cross-search algorithm for motion estimation. *IEEE Transactions on Communications*, **38** (7).
[6] Puri, A., Hang, H.M. and Schilling, D.L. (1987) An efficient block-matching algorithm for motion compensated coding. *Proceedings of the IEEE*, **ICASSP-87**, 2.4.1–25.4.4.
[7] Zhu, S. and Ma, K.-K. (2000) A new diamond search algorithm for fast block-matching motion estimation. *IEEE Transactions on Image Processing*, **9** (2), 287–290.
[8] Cheung, C.H. and Po, L.M. (2002) A Novel Cross-Diamond Search Algorithm for Fast Block Motion Estimation. *IEEE Transactions on. Circuits and Systems for Video Technology*, **12**, 1168–1177.
[9] Cheung, C.H. and Po, L.M. (2002) *A Novel Small-Cross-Diamond Search Algorithm for Fast Video Coding and Videoconferencing Applications*. Proceedings of the IEEE International Conference on Image Processing I, Rochester, pp. 681–684.
[10] Lam, C.W., Po, L.M. and Cheung, C.H. (2002) A Novel Kite-Cross-Diamond Search Algorithm for Fast Block Matching Motion Estimation. *IEEE International Symposium on Circuits and Systems*, **3**, Vancouver, pp. 729–732.
[11] Nie, Y. and Ma, K.-K. (2002) Adaptive rood pattern search for fast block-matching motion estimation. *IEEE Transactions on Image Processing*, **11** (12), 1442–1448.
[12] Zhao, H., Yu, X.B., Sun, J.H. *et al.* (2008) *An Enhanced Adaptive Rood Pattern Search Algorithm for Fast Block Matching Motion Estimation*. 2008 IEEE Congress on Image and Signal Processing, pp. 416–420.

[13] Schuster, G.M. and Katsaggelos, A.K. (1998) An optimal quad-tree-based motion estimation and motion-compensated interpolation scheme for video compression. *IEEE Transactions on Image Processing*, **7** (11), 1505–1523.

[14] Cordell, P.J. and Clarke, R.J. (1992) Low bit rate image sequence coding using spatial decomposition. *Proceedings of IEE I*, **139**, 575–581.

[15] Seferidis, V. and Ghanbari, M. (1994) Generalised block-matching motion estimation using quad-tree structured spatial decomposition. *IEE Proceedings on Vision, Image and Signal Processing*, **141** (6).

[16] Sullivan, G.J. and Baker, R.L. (1994) Efficient quadtree coding of images and video. *IEEE Transactions on Image Processing*, **3**, 327–331.

[17] Lee J. (1995) *Optimal Quadtree For Variable Block Size Motion Estimation*. Proceedings of the International Conference on Image Processing, Vol. **3**, pp. 480–483.

[18] Schuster G.M. and Katsaggelos, A.K. (1996) *An Optimal Quad-Tree-Based Motion Estimator*. Proceedings of the SPIE Conference on Digital Compression Technologies and Systems for Video Communications, Berlin, Germany, pp. 50–61.

[19] Packwood R.A., Steliaros, M.K. and Martin, G.R. (1997) *Variable Size Block Matching Motion Compensation for Object-Based Video Coding*. IEE Conference on IPA97, no. 443, pp. 56–60.

[20] Bierling, M. (1988) *Displacement Estimation by Hierarchical Block Matching*. SPIE Conference on Visual Communications and Image Processing, Cambridge, MA, pp. 942–951.

[21] Nam, K.M., Kim, J.S., Park, R.H. and Shim, Y.S. (1995) A fast hierarchical motion vector estimation algorithm using mean pyramid. *IEEE Transactions on Circuits and Systems for Video Technology*, **5**, 344–351.

[22] Chalidabhongse, J. and Kuo, C.C.J. (1997) Fast motion vector estimation using multiresolution spatio-temporal correlations. *IEEE Transactions on Circuits and Systems for Video Technology*, **7**, 477–488.

[23] Lee, J.H., Lim, K.W., Song, B.C. and Ra, J.B. (2001) A fast multi-resolution block matching algorithm and its LSI architecture for low bit-rate video coding. *IEEE Transactions on Circuits and Systems for Video Technology*, **11**, 1289–1301.

[24] Song, B.C. and Chun, K.W. (2004) Multi-resolution block matching algorithm and its VLSI architecture for fast motion estimation in an MPEG$-$2 video encoder. *IEEE Transactions on Circuits and Systems for Video Technology*, **14**, 1119–137.

[25] Lin, C.C., Lin, Y.K. and Chang, T.S. (2007) *PMRME: A Parallel Multi-Resolution Motion Estimation Algorithm and Architecture for HDTV Sized H.264 Video Coding*. Proceedings of the IEEE International Conference on Acoustics, Speech, and Signal Processing II, Honolulu, pp. 385–388.

[26] Draft ITU-T Recommendation and Final Draft International Standard of Joint Video Specification (ITU-T Rec. H.264/ISO/IEC 14 496–10 AVC), JVT-G050, 2003.

[27] Kalva, H. (2006) The H.264 video coding standard. *IEEE MultiMedia*, **13** (4), 86–90.

[28] Girod, B. (1993) Motion-compensating prediction with fractional-pel accuracy. *IEEE Transactions on Communications*, **41** (4), 604–612.

[29] Wedi, T. (2003) Motion compensation in H.264/AVC. *IEEE Transactions on Circuits and Systems for Video Technology*, **13** (7), 577–586.

[30] Y.-W. Huang *et al.* (2003) *Analysis and Reduction of Reference Frames for Motion Estimation in MPEG-4 AVC/JVT/H.264*. Proceedings of the IEEE ICASSP, pp. 145–148, Hong Kong.

[31] Su, Y. and Sun, M.-T. (March 2006) Fast Multiple Reference Frame Motion Estimation For H.264/AVC.

[32] Bross, B., Han, W.-J., Ohm, J.-R., *et al.* (July 2003) High efficiency video coding (HEVC) text specification draft 10 (for FDIS & Consent), Geneva, Switzerland, document JCTVC-L1003 of JCT-VC.

[33] Marpe, D., Shwarz, H., Bosse, S. *et al.* (2010) Video compression using nested quadtree structures, leaf merging, and improved techniques for motion representation and entropy coding, *IEEE Transactions on Circuits and Systems*, **20**(12), 1676–1687.

[34] Song, L., Tang, X., Zhang, W. *et al.* (2013) *The SJTU 4K Video Sequence Dataset*. Fifth International Workshop on Quality of Multimedia Experience (QoMEX2013), Klagenfurt, Austria, July 3–5.

[35] Kuehenne, H., Jhuang, H., Garrote, E. *et al.* (2011) *HDMB: A Large Video Database For Human Motion Recognition*. ICCV, pp. 2556–2563.

[36] Seshadrinathan, K., Soundararajan, R., Bovik, A.C. and Cormack, L.K. (2010) Study of subjective and objective quality assessment of video. *IEEE Transactions on Image Processing*, **19** (6), 1427–1441.

[37] Barron, J.L., Fleet, D. and Beauchemin, S. (1994) Performance of optical flow techniques. *International Journal of Computer Vision*, **12** (1), 43–77.

[38] Winkler, S. (2005) *Digital Video Quality – Vision Models and Metrics*, John Wiley & Sons.

[39] Wang, Z., Bovik, A., Sheikh, H. and Simoncelli, E. (2004) Image quality assessment: from error measurement to structural similarity. *IEEE Transactions on Image Processing*, **13** (4), 600–612.

[40] Sheikh, R., Sabir, M. and Bovik, A.C. (2006) A statistical evaluation of recent full reference image quality assessment algorithms. *IEEE Transactions on Image Processing*, **15** (11), 3440–3451.

[41] Wang, Z. and Bovik, A.C. (2002) A universal image quality index. *IEEE Signal Processing Letters*, **9** (3), 81–84.

[42] Wang, Z., Bovik, A.C. and Lu, L. (May 2002) *Why is Image Quality Assessment So Difficult?* Proceedings of the IEEE International Conference on Acoustics, Speech, and Signal Processing, vol. **4**, pp. 3313–3316.

[43] Wang, Z. (2001) Rate scalable foveated image and video communications. PhD Thesis. Department of ECE, The University of Texas at Austin, December.

[44] Wang, Z., Ligang, L. and Bovik, A.C. (2004) Video quality assessment based on structural distortion measurement. *Signal Processing: Image Communication*, **19** (2), 121–132.

[45] Wang, Z. and Simoncelli, E.P. (2005) *Reduced-Reference Image Quality Assessment Using A Wavelet-Domain Natural Image Statistic Model*. Human vision and Electronic Imaging, pp. 149–159, March.

[46] Martini, M.G., Mazzotti, M., Lamy-Bergot, C. *et al.* (2007) Content adaptive network aware joint optimization of wireless video transmission. *IEEE Communications Magazine*, **45** (1), 84–90.

[47] Marr, D. and Hildreth, E. (1980) *Theory of Edge Detection*. Proceedings of the Royal Society of London, Series B.

[48] Zhang, M. and Mou, X. (2008) *A Psychovisual Image Quality Metric Based on Multi-Scale Structure Similarity*. Proceedings of the IEEE International Conference on Image Processing (ICIP), San Diego, CA, October, pp. 381–384.

[49] Martini, M.G., Hewage, C.T.E.R. and Villarini, B. (2012) Image quality assessment based on edge preservation. *Signal Processing: Image Communication*, **27**, 875–882.

[50] Pinson, M.H. and Wolf, S. (2004) A new standardized method for objectively measuring video quality. *IEEE Transactions on Broadcasting*, **50** (3), 312–322.

[51] Wolf, S. and Pinson, M. (2009) *In-Service Performance Metrics For Mpeg-2 Video Systems*. Proceedings of the Made to Measure 98 – Measurement Techniques of the Digital Age Technical Seminar, International Academy of Broadcasting (IAB), ITU and Technical University of Braunschweig, Montreux, Switzerland, November.

[52] Wolf, S. and Pinson, M.H. (2005) *Low Bandwidth Reduced Reference Video Quality Monitoring System*. Video Processing and Quality Metrics for Consumer Electronics.

[53] Gunawan, I. and Ghanbari, M. (2008) Reduced-reference video quality assessment using discriminative local harmonic strength with motion consideration. *IEEE Transactions on Circuits and Systems for Video Technology*, **18**.

[54] Final report from the video quality experts group on the validation of objective models of video quality assessment, phase II, Video quality expert group, August 2003.

[55] Carnec, M., Callet, P.L. and Barba, D. (2008) Objective quality assessment of color images based on a generic perceptual reduced reference. *Signal Processing: Image Communication*, **23**.

[56] Engelke, U., Kusuma, M., Zepernick, H. and Caldera, M. (2009) Objective quality assessment of color images based on a generic perceptual reduced reference. *Signal Processing: Image Communication*, **24**.

[57] Li, Q. and Wang, Z. (2009) Reduced-reference image quality assessment using divisive normalization-based image representation. *IEEE Journal of Selected Topics in Signal Processing*, **3**.

[58] Fechter, F. (2000) Zur Beurteilung der Qualität komprimierter Bildfolgen. Habil., TU-Braunschweig, Institut für Nachrichtentechnik, ISBN 3-18-363910-6, VDI Verlag, Düsseldorf.

[59] Yang, F. *et al.* (2005) A novel objective no-reference metric for digital video quality assessment. *IEEE Signal Processing Letters*, **12** (10).

[60] Ha, H., Park, J., Lee, S. and Bovik, A.C. (2010) Perceptually Unequal Packet Loss Protection by Weighting Saliency and Error Propagation. *IEEE Transactions on Circuits and Systems for Video Technology*, **20** (9), 1187–1199.

3

Registration for Motion Estimation and Object Tracking

3.1 Introduction

Motion estimation from image sequences is one of the cornerstones in computer vision and robotics. The extraction and use of motion cues are extensively employed as a low-level feature in many multidisciplinary applications such as object detection, object tracking, vehicle navigation and activity recognition.

This diversity in the application field has been translated onto an even wider variety of motion representations: from vectors to silhouettes. Luckily, all of them can be seen as registration problem.

Motion estimation in video analytics can be understood as a temporal registration between the frames composing the video. Since registration is the process of finding the spatial transformation that best maps an image onto another image or model, motion during a sequence can be detected by comparing both images into the same space. Registration is, therefore, a general framework that relates most motion estimation techniques.

In this framework, motion estimation in static sequences, that is, where the camera is not moving, such as background subtraction, may be considered a special case of the general framework. On the one hand, temporal registration of consecutive frames allows estimating the relative motion [1, 2]. On the other hand, by using a background model as reference, image registration may be used for solving the problem of estimation. Under these conditions, registrations became trivial, and motion is simply estimated as a subtraction of pixels [3] or, more robustly, as a distance to a Gaussian model [4]. Several techniques for updating the reference model, such as median filters, ensure the validity of the approach and the accuracy of the results.

In scenarios where the camera is moving, the potential of image registration for motion estimation is clearly stated. Thus, by finding the transformation that relates

Image, Video & 3D Data Registration: Medical, Satellite & Video Processing Applications with Quality Metrics,
First Edition. Vasileios Argyriou, Jesus Martinez Del Rincon, Barbara Villarini and Alexis Roche.
© 2015 John Wiley & Sons, Ltd. Published 2015 by John Wiley & Sons, Ltd.

two consecutive frames, both the global motion of the camera and the relative motion of an object of interest within the field of view can be estimated simultaneously. This approach has been extensively used on cameras mounted on moving vehicles [5] or PTZ cameras [6], where the pan tilt and zoom can be extracted by using registration.

3.1.1   Mathematical Notation

In registration terms, the geometric transform that relates two video frames I_1 and I_2 or a frame I and a background image B can be expressed as

$$I_2(\hat{x}, \hat{y}) = I_2(f(x, y), g(x, y)) = I_1(x, y) \tag{3.1}$$

or

$$I(\hat{x}, \hat{y}) = I(f(x, y), g(x, y)) = B(x, y) \tag{3.2}$$

Obtaining the transformation functions f and g allows discovering the relationship between the two frames so that the data initially located at position (x, y) in the initial frame is now located at $(\hat{x}, \hat{y})$ in the new frame. Estimating the transformation functions is, therefore, estimating the motion at every single possible location. This estimation can be analytical in some scenarios and applications, for example, the static camera scenario, or numerically defined at every location when more complex scenarios are considered, for example, several non-linear motions (camera and object of interest) taking place simultaneously.

The most common approach to model this transformation is as a vector field:

$$f(x, y) = x + \Delta_x(x, y); g(x, y) = y + \Delta_y(x, y) \tag{3.3}$$

where Δ defines the horizontal and vertical components of the vector representing the motion of a given location. Depending on the computation cost requirements of our application problem, each location (x, y) will be at block [7] or at pixel level [8], being also possible to have a sub-pixel estimation [9].

Since a video is a 3D entity, the previous mathematical notation can be easily extended to 3D for fully modelling motion estimation on video registration, so that

$$\hat{I}_1(\hat{x}, \hat{y}, t + \Delta_t) = I_1(x + \Delta_x(x, y), y + \Delta_y(x, y), t)] \tag{3.4}$$

where $\hat{I}_1$ is an approximation of the true frame I_2 at $t + \Delta_t$ obtained by transforming the previous true frame I_1 at t. By minimizing the difference between the approximated and the true frame at $t + \Delta_t$, the motion can be estimated as follows:

$$\min_{\Delta_x, \Delta_y} (\|I_2 - \hat{I}_1\|^2) \tag{3.5}$$

3.2   Optical Flow

Optical flow is one of the most commonly used techniques for calculating the vector field in motion estimation applications. This method attempts to calculate the local vector field based on brightness changes between pixels at images I_1 and I_2. The methodology works under the assumption that these brightness changes are originated by absolute movements of the objects within the field of view. Many variation of optical flow can be found in the literature, up to the point that this terminology refers to a family of methods than to a specific algorithm. As common characteristic, these methods approximate the vector field by matching pixels with the same brightness or intensity.

Given a position (x, y) at time t, optical flow attempts to locate the new location $(\hat{x}, \hat{y}) = (x + \Delta_x(x, y), y + \Delta_y x, y))$ at time $t + \Delta_t$. Thus, optical flow can be undertaken as described in Equation (3.4). By ensuring the preservation of eq. 3.4, also known as brightness constraint, optical flow is able to approximate the real 3D motion as a 2D projection, which results in the vector field.

Assuming small temporal and spatial displacements, the brightness constraint is approximated as a Taylor series, so that

$$I(x, y, t) = I(x, y, t) + \Delta_x(x, y)\frac{\partial I(x, y, t)}{\partial x} + \Delta_y(x, y)\frac{\partial I(x, y, t)}{\partial y} + \Delta_t\frac{\partial I(x, y, t)}{\partial t} \qquad (3.6)$$

Note that higher orders of magnitude can also be considered in this approximation.

Since the temporal displacement is usually fixed to consecutive frames $\Delta_t = 1$, the vector fields are extracted by solving the equation

$$0 = \Delta_x(x, y)\frac{\partial I(x, y, t)}{\partial x} + \Delta_y(x, y)\frac{\partial I(x, y, t)}{\partial y} + \frac{\partial I(x, y, t)}{\partial t} \qquad (3.7)$$

A simpler notation can be used for the partial derivatives given their discrete nature, obtaining the simplified equation:

$$\Delta_x I_x + \Delta_y I_y = -I_x \qquad (3.8)$$

Then, by minimizing the following term, the solution could be found.

$$\min_{\Delta_x, \Delta_y}(\| \Delta_x I_x + \Delta_y I_y + I_t\|^2) \qquad (3.9)$$

However, Equation (3.9) is undetermined, that is, more than one solution can be defined. In fact, all the points (Δ_x, Δ_y) within the straight line defined by Equation (3.8) are plausible solutions. This problem is known as aperture problem and it derives from the attempt of estimating the motion using only local information. Several conditions or restrictions should be applied to estimate a unique solution. These conditions generally relate to the smoothness of the vector field, either locally, globally or as an hybrid approach.

3.2.1 Horn–Schunk Method

The Horn–Schunk method [10] is a classical approach that introduces a global smoothness constraint to the aperture problem. Under this constraint, all the image points are equally considered, and therefore, motion estimation can be understood as a parallel projection model. This is established by minimizing the smoothness constraint:

$$E_s(\Delta_x, \Delta_y) = \int\int (\Delta_{xx}^2(x, y) + \Delta_{xy}^2(x, y) + \Delta_{yx}^2(x, y) + \Delta_{yy}^2(x, y))dx\, dy \quad (3.10)$$

which is added to the brightness constraint derived from eq. 3.8:

$$E_b(\Delta_x, \Delta_y) = \int\int (\Delta_x(x, y)I_x(x, y, t) + \Delta_y(x, y)I_y(x, y, t) + I_t(x, y, t))^2 dx\, dy \quad (3.11)$$

obtaining

$$E_{HS}(\Delta_x, \Delta_y) = E_b(\Delta_x, \Delta_y) + \alpha E_s(\Delta_x, \Delta_y) \quad (3.12)$$

where α is a regularization parameter so that a higher value leads to smoother flow.

By minimizing this function, an iteration-type solution can be found, so that

$$\Delta_x^{k+1}(x, y) =$$
$$\bar{\Delta}_x^k(x, y) - \frac{I_x(x, y, t)\bar{\Delta}_x^k(x, y) + I_y(x, y, t)\bar{\Delta}_y^k(x, y) + I_t(x, y, t)}{\alpha + I_x^2(x, y, t) + I_y^2(x, y, t)} I_x(x, y, t) \quad (3.13)$$

$$\Delta_y^{k+1}(x, y) =$$
$$\bar{\Delta}_y^k(x, y) - \frac{I_x(x, y, t)\bar{\Delta}_x^k(x, y) + I_y(x, y, t)\bar{\Delta}_y^k(x, y) + I_t(x, y, t)}{\alpha + I_x^2(x, y, t) + I_y^2(x, y, t)} I_y(x, y, t) \quad (3.14)$$

being k the iteration number.

This method has the advantage of producing a smooth and highly dense optical flow. As main drawback, it is quite sensible to noise and computationally more expensive than other methodologies.

3.2.2 Lukas–Kanade Method

The Lukas–Kanade method [8] also introduces the concept of smoothness but in a different way from the previous method. The smoothness constraint is restricted here within a neighbourhood, so it is assumed that the vector flow is almost constant in each given pixel's neighbourhood. This constraint allows solving the ambiguity of the optical flow equation 3.8 by applying least squares within the neighbourhood.

A neighbourhood Nh around a given pixel (x, y) is defined comprising all the pixels within a surrounding window of $2C + 1$ and $2R + 1$ elements in horizontal and vertical components, respectively, so $N = [(x - C, y - R) \cdots (x - C, y) \cdots (x, y) \cdots (x + C, y) \cdots (x + C, y + R)]$. The optical flow equation within N must satisfy

$$\Delta_x I_x(N_n) + \Delta_y I_y(N_n) = -I_x(N_n) \quad \forall n \in N \tag{3.15}$$

This set of equation can be rewritten using matrix notation so $S \cdot \Delta = T$:

$$\begin{bmatrix} I_x(N_1) & I_y(N_1) \\ I_x(N_2) & I_y(N_2) \\ \vdots & \vdots \\ I_x(N_N) & I_y(N_N) \end{bmatrix} \begin{bmatrix} \Delta_x \\ \Delta_y \end{bmatrix} = \begin{bmatrix} -I_t(N_1) \\ -I_t(N_2) \\ \vdots \\ -I_t(N_N) \end{bmatrix} \tag{3.16}$$

By selecting an appropriate neighbourhood size, the system is over-determined since it contains more equations than unknown variables. Least-square method can therefore be applied to solve it locally, so

$$S^T S \Delta = S^T T \quad \Rightarrow \quad \Delta = (S^T S)^{-1} S^T T \tag{3.17}$$

given the final local solution

$$\begin{bmatrix} \Delta_x \\ \Delta_y \end{bmatrix} = \begin{bmatrix} \sum_n I_x(N_n)^2 & \sum_n I_x(N_n)I_y(N_n) \\ \sum_n I_y(N_n)I_x(N_n) & \sum_n I_y(N_n)^2 \end{bmatrix}^{-1} \begin{bmatrix} -\sum_n I_x(N_n)I_t(N_n) \\ -\sum_n I_y(N_n)I_t(N_n) \end{bmatrix} \tag{3.18}$$

On the contrary to Horn–Schunk, LK method is less affected by noise but it cannot provide motion estimation within uniform areas due to its inherent local nature.

3.2.3 Applications of Optical Flow for Motion Estimation

Optical flow has many application fields, including image stabilization, motion-based data compression, 3D scene reconstruction, dynamical processes analysis, autonomous navigation or activity recognition. Among them, motion estimation for object/human detection and tracking is the most relevant for us in this chapter.

It is important, however, to understand that optical flow only approximates the real motion field. The real 3D motion field is almost impossible to estimate unambiguously [11] due to the presence of multiple motion sources, occlusions and self-occlusions, projective geometry, illumination changes, noise or the previously mentioned aperture problem. However, it can be related to transactional and rotational velocities of major rigid surfaces on the scene, given a rough approximation, enough for most applications.

3.2.3.1   Optical Flow-Based Object Detector

In application where the camera is static, the motion field is directly used to estimate the motion of the objects of interest [1, 2].

A simple example is depicted in Figure 3.1. Once the image has been captured and converted to greyscale or intensity, a chosen optical flow algorithm is applied. The motion field is then segmented and clusterized into objects by applying median filter to remove noise and thresholding to detect significative moving pixels. Morphological operators open and close help to group moving pixels into compact blobs, getting ride or holes and other imperfections. A final blob analysis based on connectivity allows segmenting the moving object of interest. Results of this system are shown in Figure 3.2.

3.2.3.2   Body Part Detection and Pose Estimation

Given its good performance and high resolution, optical flow has also been applied for estimating the relative motion or objects and body parts [12–15], deformable objects

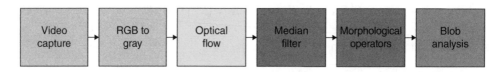

Figure 3.1   Flow diagram of an OF-based object detector [2]

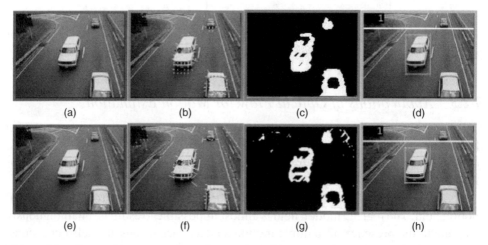

Figure 3.2   Car tracking system based on optical flow estimation. a&e)original image containing the objects of interest, b) motion field estimation using Horn–Schunk optical flow, f) motion field estimation using Lukas–Kanade optical flow, c&g)foreground segmentation, d&h) object tracking

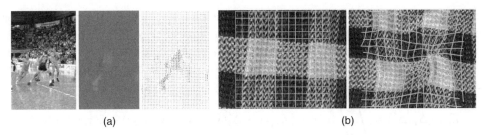

(a) (b)

Figure 3.3 (a) Large displacement optical flow used for body part segmentation and pose Estimation; (b) deformable object tracking using optical flow constraints

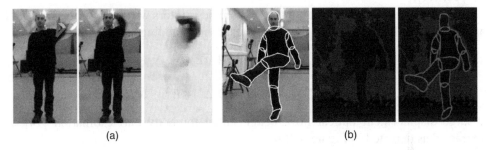

(a) (b)

Figure 3.4 Human pose estimation using optical flow features. (a) Articulated pose tracking using optical flow for dealing with self-occlusion; (b) upper-body human pose estimation using flowing puppets, consisting on articulated shape models with dense optical flow [20]

[16] (see Fig. 3.3) and articulated chains [13, 14] (see Fig. 3.4), and being a component of the last-generation part-based detectors.

Human articulated motion is a challenging problem, where small body parts at the end of the articulated chain can move very fast in comparison with the previous structure. It is in this context where optical flow can provide discriminative vectors able to individualize the body parts [17, 18]. For instance, coarse-to-fine schemas [19] have been applied to fully exploit the articulated definition of the problem, where a bigger structure gives a first indication of the following level. However, conventional optical flow techniques have been reported unreliable for big and violent displacements, which are critical in the case of the hands or lower limbs. To solve this issue, recent approaches [12] have approached a calculation to optical flow fields robust to large displacements by means of combining conventional optical flow with featured-based motion estimation techniques. This combination is performed by introducing two new constraints to the previously described brightness 3.11 and smoothness 3.10 constraints. First, a matching constraint integrates the point correspondences from feature descriptor matching with the optical flow framework:

$$E_{\text{match}}(\Delta_x, \Delta_y, \Delta'_x, \Delta'_y) =$$

$$\int \int (\delta(x, y)\rho(x, y)((\Delta_x(x, y) - \Delta'_x(x, y))^2 + (\Delta_y(x, y) - \Delta'_y(x, y))^2))dx\, dy \quad (3.19)$$

where $\Delta'(x, y)$ denotes the correspondence vectors obtained by feature descriptor matching at some points (x, y), δ is a function equal to 1 or 0 if each descriptor is available or not in frame t, and ρ contains the matching score of each correspondence;

A second constraint is needed to perform the matching between feature descriptors during the optimization

$$E_{\text{desc}}(\Delta'_x, \Delta'_y) = \int \int (\delta(x, y)|f_{t+1}(x + \Delta'_x(x, y), y + \Delta'_y(x, y)) - f_t(x, y)|^2)dx\, dy$$

(3.20)

where f_t and f_{t+1} denote the (sparse) fields of feature vectors in frame t and $t + 1$, respectively.

Feature-based motion estimation and feature descriptors will be described in detail in Section 3.3. By minimizing the energy resulting in adding all the constrains

$$E(\Delta_x, \Delta_y) = E_b(\Delta_x, \Delta_y) + \alpha E_s(\Delta_x, \Delta_y) +$$

(3.21)

$$+ \beta E_{\text{match}}(\Delta_x, \Delta_y, \Delta'_x, \Delta'_y) + E_{\text{desc}}(\Delta'_x, \Delta'_y)$$

a large displacement optical flow, better suited for body part estimation [14], is extracted as depicted in Figure 3.3(a).

3.2.3.3 Simultaneous Estimation of Global and Relative Motions

In case of dynamic scene perception where the camera parameters are unknown, the vector field makes possible identifying, simultaneously, the camera motion and the relative motion of the tracked object [21]. This is a common situation for autonomous vehicles, assisted driving or surveillance sequences recorded from moving vehicles or PTZ cameras. For those dynamic scene applications without objects of interest, the estimated global motion can be used to stabilize the image and simplify the postprocessing [22] or to reconstruct the scenario in 3D [23].

A common and robust approach to segment the dense optical flow in background and foreground layers consists of describing the vector field within a Bayesian framework [24, 25]. In this framework, background and foreground can be estimated by applying expectation-maximization [21], mixture models [4, 24] or fast flood filling [26]. Thus, Yalcin *et al.* [21] formulated the simultaneous estimation of the background B_t, foreground and motion estimations (Δ'_x, Δ'_y) by maximizing the Bayesian probability:

$$\arg \max_{B_t, (\Delta'_x, \Delta'_y)} P(B_t, \Delta^t_x, \Delta^t_y | B^S_{t-1}, I_t, I^S_{t-1}, \Delta^{t-1}_x, \Delta^{t-1}_y|)$$

(3.22)

where the superindex S means that the previous image and background should be stabilized in order to cope with big displacements. The full system is described in Figure 3.5 and results of the approach are shown in Figure 3.6.

On those application where the global motion can be estimated, such as robotics through egomotion or PTZ camera through user input parameters, this information

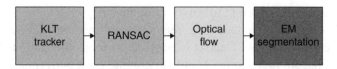

Figure 3.5 Diagram of vehicle detection in surveillance airborne video described in [21]. Images are first stabilized by detecting invariant features, which are used to estimate the affine transformation using RANSAC. Then, dense optical flow is estimated and segmented using EM clustering

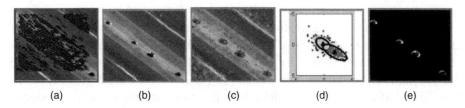

(a)	(b)	(c)	(d)	(e)

Figure 3.6 (a) KLT features used for stabilization. (b) Stabilized image. (c) Optical flow clusterized according to EM. (d) EM clusters. (e) Resulting foreground segmentation. Images from Ref. [21]

is often fed back to the segmentation process in order to enhance the moving object detection [5, 26]

3.3 Efficient Discriminative Features for Motion Estimation

Optical flow aims to estimate the motion vectors directly from low-level information, that is, pixel value or intensity. This and other methods are called direct or global-based motion estimation methods, since they do not try to understand or segment the image during processing but operating at a very low level. Since all the pixels are equally considered, heavy processing is needed. In order to mitigate the problem, subsampling both spatially and temporally has been applied, such as targeted by the block matching algorithm [7], where pixels are grouped into blocks to reduce computation. New-generation computers and parallel programming devices have allowed the increasing application of direct methods in real time, but they are still banned for implementation into embedded devices.

On the other hand, a second family of methods, called indirect or feature based, allow us to estimate the motion vectors from higher level features. By detecting more relevant features, a simple understanding of the structure of the image, including the surfaces and objects of interest, is achieved. This reduces the data size and therefore the computation during the optimization process. These features are also more robust and invariant, and less sensible to illumination changes and geometrical transformations since they are associated with edges, corners or textures.

While global methods attempt to solve simultaneously both the motion estimation and the field of view correspondence, feature-based methods generally divide the problem in these two stages. Thus, it is possible to focus the computational power on those areas that are first spotted as able to provide good correspondence. Torr and Zisserman advocated for this dual-stage schema [27]. Other authors have suggested hybrid schemas to combine the advantages of both methodologies [12, 21].

3.3.1 Invariant Features

Extensive research has been done on increasingly better and more efficient features for motion estimation. These feature extraction methods cover from the simple edge extraction based on Canny [28] or Laplacian of Gaussians [29] to the more advance and efficient features, passing by traditional corner detectors such as Harris [30], Lowe [31] and bank of filters such as Haar [32], Gabor [33], wavelets [34] or steerable filters [35]. Under the motto 'the more invariant the features, the better motion estimation', extensive research has been conducted in the pursuit of the perfect feature descriptor.

Scale-invariant feature transform (SIFT) was proposed by Lowe [31], and it is accepted as one of the most invariant and versatile feature detectors. Each feature is represented as a 3D histogram of gradients, where gradient location and orientation are weighted using its magnitude. This descriptor is integrated within a scale-invariant detector. This methodology is invariant to translation, scale and rotation, and robust against illumination changes and geometric distortion produce for the 3D to 2D projection.

Many SIFT variations [36, 37] have been evolved from the original algorithm, trying to overcome some of its limitations but keeping the strengths. For instance, principal component analysis SIFT (PCA-SIFT) [38] aims to provide similar accuracy than SIFT while reducing the complexity. This is achieved by reducing the dimensionality, that is, the length of the descriptor vector, one order of magnitude using PCA.

Gradient location and orientation histogram (GLOH), is also an extension of SIFT where a log-polar location grid is used to register the radial gradient directions. It has been proved [39] as one of the best descriptors. Similar to PCA-SIFT, it makes use of PCA to reduce the high dimensionality of the feature vector and improve its efficiency.

Speeded up robust features (SURF) [40] is a high-performance feature detector. It was partially inspired on SIFT and was developed to achieve similar performance than SIFT but with significantly less complexity and required processing time. The feature descriptor is composed of the distribution of Haar wavelet responses around the point of interest. SURF uses a fast Hessian matrix-based detector, which can be efficiently computed using integral images. While clearly outperforming SIFT on efficiency, it has not been proved to overcome it when time is not a constraint [41]. As main conceptual differences, SURF is based on Haar wavelets instead of derivative approximations in an image pyramid, it employs the Hessian matrix instead of the

Laplacian operator to extract the interest points, and the feature descriptors are sums of first-order derivatives instead of histograms of directions.

Histogram of oriented gradients (HOG) was initially proposed by Dalal and Triggs [42]. HOG was originally designed as a general object recognition and extensively used in human motion estimation [42, 43]. However, its good performance and solid base have widened its application to other fields such as image registration or pose estimation. Both SIFT and HOG make use of histogram of gradient orientations, having the similar underlying idea, due to its invariant properties. The main differences are the usage of uniform sampling, bigger cells (since it aims to recognize regions more than key points) and fine orientation binning. HOG is therefore considered a regional image descriptor more than a local image descriptor such as SIFT or SURF. In addition, in contrast to SIFT, it is not invariant to rotation since the descriptor is not normalized regarding the orientation.

Many other features can be found in the literature. Particularly interesting due to their high performance are those methods using binary strings. Local binary patterns (LBP) [44] outstands due to its simplicity to capture and describe texture by using binary numbers. Its main properties are the discriminative power and robustness against illumination changes and its computational efficiency. Since it is based on texture instead of gradients, they have been extensively used when dealing with natural images and deformable objects. Current state-of-art descriptors Binary Robust Invariant Scalable Keypoints, BRISK, [45] and Fast Retina Keypoint descriptor (FREAK) [46] also use binary strings to increase its performance almost an order of magnitude regarding SURF while maintaining and increasing the robustness and invariance. Both descriptors use a circular sampling pattern to characterize the keypoint. Other interesting binary descriptors are ORB [47] and BRIEF [48] (Figure 3.7).

Recent comparisons between SURF, BRISK and FREAK have been performed in order to determine the best feature extractor in the context of SLAM and human detection [49, 50]. Conclusions vary depending on application and dataset. In general terms, it can be stated that SURF remains competitive but is slower than BRISK and FREAK, whose performance is slightly less accurate but have real potential in embedded applications.

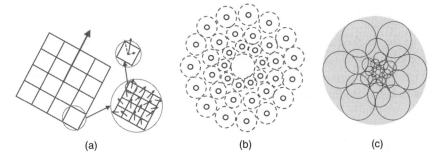

(a)   (b)   (c)

Figure 3.7 Sampling patterns belonging to the feature-based descriptors: (a) SIFT and SURF [40]; (b) BRISK [45]; and (c) FREAK [46]

3.3.2   Optimization Stage

Once the feature descriptors have been calculated and extracted, matching between the feature descriptor sets belonging to consecutive images must be performed in order to estimate the global and relative motion parameters. Since descriptor calculation errors are uncorrelated between features [27], it is valid to assume that they are statistically independent during the estimation.

Random sample consensus, or RANSAC, is the most commonly used optimization method to estimate motion parameters from a set of noisy data. This iterative method models the presence of outliers during the estimation, which suits the registration problem where errors in feature extraction or presence of non-affine transformation are common. This complements the previous stage and increase the robustness of the framework. As main drawback, RANSAC may not be able to estimate the optimal solution if the number of iteration is limited and/or the outliers surpass the inliers.

Other similar methodologies include regression, spectral and graph matching and Hough transform [51]. While these methods work well for feature-based methodologies where the number of keypoints is small, the suffer from complexity that prevent their usage in global approaches or when the non-linearity of the data is big.

3.4   Object Tracking

Feature-based tracking is one of most frequent approach in object and human tracking. This is due to its invariance to illumination changes (compared to colour-based tracking), resistance to occlusion and possibility to track non-rigid deformable objects (compared to template-based and contour tracking).

A perfect example of feature-based tracker is the Kanade-Lucas–Tomasi (KLT) tracker [52]. This algorithm has been used as integral part of many human trackers and surveillance systems, such as [53, 54]. Other similar feature trackers similar to KLT are [55–57].

When object tracking is the goal of our application, previous direct and indirect motion estimation methods are directly applicable. However, these methodologies estimate simultaneously the global motion due to the camera and the motion of the object of interest. As a consequence, the estimated motion will only represent the motion of the tracked object if the camera is assumed static or the global motion is subtracted. This is usually performed by considering the object of interest as an outlier [21], since the object occupy generally a smaller area of the field of view, which allow us to calculate the global background motion as the average estimated motion.

3.4.1   KLT Tracking

KLT formulates object tracking as a translational registration problem. Under this formulation, the displacement Δ_x of the tracked feature $I(\mathbf{x})$ from the previous location

$\mathbf{x} = (x, y)$ between two images I_1 and I_2 can be approximated as

$$\hat{I}_1(\mathbf{x}) \approx \frac{I_1(+\Delta_\mathbf{x}) - I_1(\mathbf{x})}{\Delta_\mathbf{x}} = \frac{I_2(\mathbf{x}) - I_1(\mathbf{x})}{\Delta_\mathbf{x}} \qquad (3.23)$$

so that

$$I_1(\mathbf{x} + \Delta_\mathbf{x}) \approx I_1(\mathbf{x}) + \Delta_\mathbf{x}\hat{I}_1(\mathbf{x}) \qquad (3.24)$$

The estimated displacement $\Delta_\mathbf{x}$ is that which minimizes the L_2-norm of the difference between the images

$$E = \sum_\mathbf{x} \left[I_1(\mathbf{x} + \Delta_\mathbf{x}) - I_2(\mathbf{x}) \right]^2 \approx \sum_\mathbf{x} \left[I_1(\mathbf{x}) + \Delta_\mathbf{x}\hat{I}_1(\mathbf{x}) - I_2(\mathbf{x}) \right]^2 \qquad (3.25)$$

This error can be minimized by partially deriving it and setting it to zero, so that

$$0 = \frac{\partial E}{\partial \Delta_\mathbf{x}} \approx \frac{\partial}{\partial \Delta_\mathbf{x}} \sum_\mathbf{x} \left[I_1(\mathbf{x}) + \Delta_\mathbf{x}\hat{I}_1(\mathbf{x}) - I_2(\mathbf{x}) \right]^2 \qquad (3.26)$$

$$= \sum_\mathbf{x} 2\hat{I}_1(\mathbf{x}) \left[I_1(\mathbf{x}) + \Delta_\mathbf{x}\hat{I}_1(\mathbf{x}) - I_2(\mathbf{x}) \right] \qquad (3.27)$$

giving the following estimation of the displacement:

$$\Delta_\mathbf{x} \approx \frac{\sum_\mathbf{x} \hat{I}_1(\mathbf{x})[I_2(\mathbf{x}) - I_1(\mathbf{x})]}{\sum_\mathbf{x} \hat{I}_1(\mathbf{x})^2} \qquad (3.28)$$

Since the estimation of I_1 is dependent on the estimation of $\Delta_\mathbf{x}$, this procedure is applied iteratively until convergence or maximum number of iterations is reached. This iterative process can be included in the formulation as

$$\Delta_\mathbf{x}^0 = 0 \qquad (3.29)$$

$$\Delta_\mathbf{x}^{k+1} = \Delta_\mathbf{x}^k + \frac{\sum_\mathbf{x} w(\mathbf{x})\hat{I}_1(\mathbf{x} + \Delta_\mathbf{x}^k) \left[I_2(\mathbf{x}) - I_1(\mathbf{x} + \Delta_\mathbf{x}^k) \right]}{\sum_\mathbf{x} w(\mathbf{x})\hat{I}_1(\mathbf{x} + \Delta_\mathbf{x}^k)^2} \qquad (3.30)$$

being k the number of the iteration and $w(\mathbf{x})$ a weighting function that measures the contribution of each estimate at a different given location $\mathbf{x}$.

This approximation is accurate if the displacement between two images is not too large (so the approximation of the first-order Taylor series is true), and the tracked object has representative and invariant keypoints. This last limitation suggests us the usage of invariant feature-based keypoints, instead of optical flow at any pixel or other global approached [58]. On the other hand, KLT allows us to track deformable objects since the object is not to reduce a unique entity but as a set of interrelated components or descriptors, as far as the deformation still keeps some keypoints from one frame to another. An example of KLT tracker for face tracking is shown in Figure 3.8.

Figure 3.8 Face location of an orientation tracking using KLT and minimum Eigenvalue features [58, 59]

3.4.2   Motion Filtering

In practice, feature-based estimation is prone to errors and can be heavily affected by outliers. Under these circumstances, the matching between images can fail leading to divergence or jittering. In order to avoid these undesirable effects, filtering theory provides a Bayesian framework where the initial motion estimation is refined according to prior and posterior information. This framework can be applied to each individual feature or to the object as a whole, by previously grouping features into objects. The grouping can be done by applying clustering, templates or simple averaging.

Bayesian filtering algorithms estimate the state $\mathbf{x}$ by predicting first the prior probability of $\mathbf{x}$ according to a motion model, and correcting later according to the observational data $\mathbf{z}$. This is condensed in the following equation:

$$p(\mathbf{x}_t|\mathbf{z}_t) = p(\mathbf{z}_t|\mathbf{x}_t) \int p(\mathbf{x}_t|\mathbf{x}_{t-1})p(\mathbf{x}_{t-1}|\mathbf{z}_{1:t-1})d\mathbf{x} \tag{3.31}$$

where $p(\mathbf{x}_{t-1}|\mathbf{z}_{1:t-1})$ is the previous estimation, $p(\mathbf{x}_t|\mathbf{x}_{t-1})$ is the motion model and $p(\mathbf{z}_t|\mathbf{x}_t)$ is defined by the observation.

This general framework can be exploited by using motion estimations given by the previous direct and indirect methods as the observation $\mathbf{z}$. With that aim, it is required to calculate a suitable function f able to approximate the probability $p(\mathbf{z}_t|\mathbf{x}_t)$. Different function can be applied, such as the L_2-norm, the SSD, the NCC or any other distances discussed in previous chapters, in the case of optical flow, or the matching score between feature descriptors, in the case of feature-based methods. For histogram-based descriptors such as HOG, SIFT and SURF, Bhattacharyya distance is commonly employed. After defining the function f, the observation probability can be described as

$$p(\mathbf{z}_t|\mathbf{x}_t) \propto \frac{1}{\sigma\sqrt{2\pi}} \exp\left(\frac{-f(D_t, D_{t-1})^2}{2\sigma^2}\right) \tag{3.32}$$

being $f(D_t, D_{t-1})$ the similarity between the tracked feature descriptor D_{t-1} and the observed descriptor D_t at the predicted location.

As a result, this Bayesian formulation provides a robust environment where the motion estimation can be refined and balanced between a predicted motion model and the noisy observed data. In addition, it reduces the searching space where registration must be performed by using the prediction as a starting searching point.

Kalman filter in its numerous variations (EKF, UKF, etc.) is the most classical example of filtering that predicts and estimates the final position by balancing the coherence of the prediction given by the motion model and the observed location. It is derived by assuming Gaussianity in Equation (3.31), which reduces the previous formulation to discrete points and covariance matrices. Its simple and efficient implementation and its condition as optimal estimator under linear conditions are responsible of its ubiquity for motion estimation [60, 61].

Monte Carlo methodologies, such as particle filter, also provide a solid framework where features can be integrated and tracked without requiring exhaustive scanning of the full image [62–64]. As a noticeable difference, while Kalman filter is applied to every object or feature, MC-based methods maintain the probabilistic formulation by approximating the probabilities as a weighted sum of N random samples:

$$p(\mathbf{x}_t|\mathbf{z}_t) \approx \sum_{i=1}^{N} w_t^i \delta(\mathbf{x}_t - \mathbf{x}_t^i) \qquad (3.33)$$

where the weight of each sample is calculates as

$$w_t^i = w_{t-1}^i p(\mathbf{z}_t|\mathbf{x}_t^i) \qquad (3.34)$$

3.4.3 Multiple Object Tracking

While tracking an isolated object in video sequences is a well-consolidated problem, tracking multiple and simultaneous targets still remains a challenge. Although it could appear as an extension of the single object tracker [64], interactions and occlusions between targets are difficult to model. As example, Monte Carlo-based methodologies have been extended to include interaction function between objects such as MCMC particle filter [65] or mean-field particle filter [66]. However, they have struggled to apply in practice due to its intense computational requirements, which limits them to a few simultaneous targets due to the curse of dimensionality, and the difficulty of designing appropriate interaction models in real scenarios.

Recently, radical improvements on feature and object detection have led to the so-called tracking-by-detection, where filtering algorithms fill a secondary role in favour of the initial estimation. The problem is then primarily reduced to data association between detection on consecutive frames. This consideration does not differ much from registration. By pairing the objects/features in one image with the corresponding positions in the next frames, the displacements can be estimated and the targets tracked, discarding those outliers as objects abandoning the scene or labelling them as occluded.

Conventional data association techniques such as the Hungarian algorithm [67], multiple hypothesis tracking MHT [68] and linear and multiple assignment problems LAP and MAP [69] have been key on the tracking-by-detection algorithms. More recently, graph theory and linear programming allow dealing with the exponentially growing multiple combinations and estimating the optimal solution in a more efficient way.

3.5   Evaluating Motion Estimation and Tracking

3.5.1   Metrics for Motion Detection

Classical metrics for detection and hypothesis testing have been commonly applied for object tracking, such as type I and type II errors. These errors, described as four numbers, reveal all the possible sources of errors of the estimates regarding a known groundtruth. This pairing between estimated and real location is condensed in the following four numbers:

True positive TP: number of real locations that are correctly estimated.
False positive FP: number of estimated locations that are not real.
True Negative TN: number of non-existing locations that are not estimated, and therefore correctly discarded.
False Negative FN: number of real locations that are not estimated.

The association between an estimate and a groundtruth location is based on an ad hoc maximum distance threshold or the standard PASCAL 50% overlap criterion [65]. It is always important to remember that the groundtruth cannot be used during the estimation process at any point but at the final evaluation.

These four metrics are usually compressed into a couple of numbers in order to facilitate understanding and draw conclusions, as well as normalizing regarding the size of the testing set. Precision and recall are broadly used for this purpose, where precision being a measure of exactness or quality while recall measures completeness or quantity.

$$\text{Precision} = \frac{\text{TP}}{\text{TP} + \text{FP}} \tag{3.35}$$

$$\text{Recall} = \frac{\text{TP}}{\text{P}} = \frac{\text{TP}}{\text{TP} + \text{FN}} \tag{3.36}$$

A similar couplet is given by the sensibility and specificity, where sensitivity relates to the ability to detect a condition correctly and specificity relates to the ability to exclude it properly.

$$\text{Sensitivity} = \frac{\text{TP}}{\text{P}} = \frac{\text{TP}}{\text{TP} + \text{FN}} \tag{3.37}$$

$$\text{Specificity} = \frac{TN}{N} = \frac{TN}{FP + TN} \qquad (3.38)$$

Accuracy that is defined in similar terms as

$$\text{Accuracy} = \frac{TP + TN}{P + N} = \frac{TP + TN}{TP + FN + FP + TN} \qquad (3.39)$$

is also commonly accepted. However, since it does not reflect properly the false positive, it is suggested to be complemented by the FP metric or the Precision.

3.5.2   Metrics for Motion Tracking

These metrics are commonly applied for object detection, that is, for motion estimates resulting from individual pairing between images but without keeping temporal consistency over time. They have also been applied to tracking result, since tracking aims to improve the results given by the same metrics applied on frame-to-frame basis by means of filtering and reducing noise and discarding some of the outliers. However, these metrics were not tailored for tracking and fail to consider the full tracking problem. In particular, they do not reflect all the complexity of multitarget scenarios, where consistency over long period of frames and labelling errors and identity exchanges between targets after interactions are common occurrences that must be quantified in order to assess the boundaries of the approach.

Therefore, in order to measure the real performance of a given system in the context of tracking, two new complementary metrics have been proposed. The CLEAR MOT performance metrics [54] define two measures of tracking performance: multiple object tracker accuracy (MOTA) and multiple object tracker precision (MOTP), defined as follows:

$$\text{MOTA} = 1 - \frac{\sum FN + FP + Id_{switch}}{P} = 1 - \frac{\sum FN + FP + Id_{switch}}{TP + FN} \qquad (3.40)$$

$$\text{MOTP} = \frac{\sum e_{TP}}{\sum TP} \qquad (3.41)$$

where e_{TP} is defined as the distance or error between the groundtruth objects' locations and the associated estimated locations, normally defined by the Euclidean distance. MOTA is widely accepted as a good reflection of true tracker performance, as it measures false-positives, false-negatives and ID-switches, whereas MOTP simply measures how closely the tracker follows the groundtruth regardless of any other errors. For this reason, MOTA will be occasionally used standalone to evaluate some of the tracking tuning decisions.

3.5.3   Metrics for Efficiency

Finally, but not less importantly, every tracking methods should be evaluated in terms of efficiency or computational cost. This metric is particularly relevant if we consider that tracking is expected to be performed in real time as a component in an action recognition and classification framework. Direct measurement of these parameters is the average time in seconds for processing a frame, both normalized by the number of targets or not. However, this depends on the quality of the implementation, the hardware, the compiler and the programming language. Therefore, it may require access to the code of state-of-art methodologies or their reimplementation to ensure fair comparison. As alternative, the theoretical complexity, described by means of the O-notation, is generally accepted as a reliable way of estimating computational complexity and facilitating comparison.

3.5.4   Datasets

All previous metrics are generated by comparing the estimated locations with the groundtruth locations, which shows clearly the necessity of groundtruthing. This can be accomplished by manually annotating the test sequences with the corresponding human effort that implies. In order to mitigate, some annotation tools have been developed to semi-supervise the annotation process such as [70]. As an alternative, some initial but not convincing attempts of evaluation using synthetic data can be found in the literature.

More interesting approach is the usage of standard datasets, as identified by many evaluation projects. A variety of standard datasets and competitive calls have been made available to the scientific community. Relevant datasets are the CHIL [71], AMI [72], VACE [73], ETISEO [74], CAVIAR [75], CREDS [76], PETS [77], Oxford [65], HumanEVA [78], EEMCV [79], AVSS [80] and CLEAR [81].

Recently i-lids [82] have been released by the UK home office, and it still remains as accountable of the current state of the art due to its elevate complexity. A summary of the main properties of each dataset is given in Tables 3.1 and 3.2.

3.6   Conclusion

In this chapter, the concept of optical flow for detection and tracking has been analysed, presenting the related state-of-the-art methods and indicating their advantages and disadvantages. Feature-based motion estimation techniques have been also discussed in a similar context, as alternative for background–foreground separation and object and human tracking for video surveillance. The concepts of multitarget tracking evaluation including methods for detection recall, detection precision, number of groundtruth detections, number of correct detections, number of false detections, number of total detections, number of total tracks in groundtruth and so on were finally

Table 3.1 Properties of the main datasets on tracking and activity recognition

Dataset	Sensors	Aim	Sequences	Groundtruth	Difficulty
CHIL	4 cameras, wide-angle cameras, array microphones	Ambient intelligence	86 clips, 40–60 min each	Transcription, location, pose, gestures	Medium-High
AMIDA	Microphone, 1 camera	Ambient intelligence	20 hours	Head location, transcription	Medium-High
VACE	News clips	Entertainment	50 clips, 30 min each	Face, person position, text	Easy
ETISEO	Single and/or multi-camera	Video-Surveillance. Object detection, tracking and classification	86 clips, 5 scenarios	Camera calibration, location, contextual info	Medium
CAVIAR I/II	Wide area camera/2 cameras	Video surveillance. Person Detection, tracking and simple activity recognition (walking, wandering, leaving bags, in and out, fighting, grouping, falling)	28+26clips	Location, event annotation	Easy
CREDS	Multi-camera	Video Surveillance. Even Detection (proximity, abandoned objects, launching objects, people trap, walking on rails, falling, crossing)	12 scenarios	Location, event annotation	Medium

(continued overleaf)

Table 3.1 (*Continued*)

Dataset	Sensors	Aim	Sequences	Groundtruth	Difficulty
i-Lids	Single and/or Multi-camera	Video-Surveillance. Abandoned baggage, parked vehicle, infrared maritime sequences, doorway surveillance and sterile zone scenario, multi-camera airport	170 clips, 30–60 min each	Location, event annotation	High
Town Center	1 camera	Multi-target tracking	1 clip, 4500 frames	Location, body part annotation	Medium–high
PETS 2009	4/8 cameras	Multi-target tracking, crowd analysis	33 clips, 60–100 s each, 40 actors	Calibration, Location, event annotation	Medium–High
PETS 2007	4 cameras	Video surveillance. Loitering, luggage theft, unattended luggage	9 clips, 100–180 s each	Location, event annotation	Medium
PETS 2006	4 cameras	Video surveillance. Abandoned object	7 clips, 100–180 sec. each	Location, event annotation	Easy

Table 3.2 Properties of the main datasets on tracking and activity recognition (continue)

Dataset	Sensors	Aim	Sequences	Groundtruth	Difficulty
PETS 2003 Smart room	2 cameras, omni-directional camera	Ambient intelligence	Several clips	Camera calibration, facial expressions, gaze and gesture/action, head and eye location	Medium
PETS 2003 Football	3 cameras	Multitarget tracking	Several clips	Camera calibration, person location	Medium
PETS 2002	1 camera	Person tracking, counting, simple event recognition	Several clips	Person location, event annotation	Easy
PETS 2001	2 cameras, omni-directional camera	Multi person and vehicle tracking	10 clips, several sec. each	Calibration, location	Easy–Medium
PETS 2000	1 camera	Multiperson tracking	210 clips	Calibration, location	Easy
Hollywood I/II	Movie clips	Activity recognition (8/12 classes: answer phone, get out car, hand shake, hug person, kiss, sit down, sit up, stand up/+ drive car, eat, fight person, run)	430/2859 clips, sec–min each, 10 scenarios	Activity	High
Weizmann	1 camera	Activity Recognition (10: run, walk, skip, jack, jump, sideways, wave, salute, bend)	90 clips, 3–5 s each, 9 actors	Activity	Very Easy

(continued overleaf)

Table 3.2 (Continued)

Dataset	Sensors	Aim	Sequences	Groundtruth	Difficulty
KTH	1 camera	Activity Recognition (6: walking, jogging, running, boxing, hand waving, hand clapping)	2391 clips, 4 scenarios, 25 actors	Activity	Very Easy
UT-interaction	1 camera	Surveillance. Activity and interaction recognition.	20 videos, 2 scenarios	Event annotation	Medium
Behave	2 cameras	Video Surveillance. Multiperson tracking. Activity recognition (In group , Approach , WalkTogether, Split, Ignore, Following, Chase, Fight, RunTogether and Meet)	4 clips, each divided in more than 7 seq.	Calibration, location, event annotation	Medium
IXMAS	5 cameras	Activity Recognition (13 actions: nothing, check watch, cross arms, scratch head, sit down, get up, turn around, walk, wave, punch, kick, point, pick up, throw)	186 clips, 11 actors	Calibration, activity, silhouette	Easy–medium
HumanEVA I/II	7/4 cameras	Pose estimation. Activity recognition (walking, jogging, gesturing, fighting)	39/2 clips. 4/2 actors	Camera calibration, 3d Mocap	Medium

presented, as well as a summary of the most relevant and up-to-date datasets on video surveillance, activity recognition and entertainment.

3.7 Exercise

This exercise is focused on extracting the motion of the objects in a video sequence using optical flow techniques. First choose and implement an optical flow methodology to extract the motion in the video sequence surveillance.avi. How well does it perform? Can you visually segment the objects? In a second stage, implement an algorithm of your choice to group the motion vectors into objects. Finally, could you extract the global motion direction of each object?

References

[1] Shin, J., Kim, S., Kang, S. *et al.* (2005) Optical flow-based real-time object tracking using non-prior training active feature model. *Real-Time Imaging*, **11**, 204–218.

[2] Aslani, S. and Mahdavi-Nasab, H. (2013) Optical flow based moving object detection and tracking for traffic surveillance. *International Journal of Electrical, Robotics, Electronics and Communications Engineering*, **7** (9), 131–135.

[3] Piccardi, M. (2004) Background subtraction techniques: a review. *IEEE International Conference on Systems, Man and Cybernetics*, **4**, 3099–3104.

[4] Stauffer, C. and Grimson, W.E.L. (1999) Adaptive background mixture models for real-time tracking. *Computer Vision and Pattern Recognition*, Vol. **2**, 246–252.

[5] Braillon, C., Pradalier, C., Crowley, J.L. and Laugier, C. (2006) Real-time moving obstacle detection using optical flow models. *Intelligent Vehicles Symposium*, 2006.

[6] Doyle, D.D., Jennings, A.L. and Black, J.T. (2013) Optical flow background subtraction for real-time PTZ camera object tracking. Instrumentation and Measurement Technology Conference, 2013, pp. 866,871.

[7] Ourselin, S., Roche, A., Prima, S. and Ayache, N. (2000) Block matching: A general framework to improve robustness of rigid registration of medical images. LNCS Medical Image Computing and Computer-Assisted Intervention, 2000, pp. 557–566.

[8] Lucas, B.D. and Kanade, T. (1981) An iterative image registration technique with an application to stereo vision. Proceedings of Imaging Understanding Workshop, 1981, pp. 121–130.

[9] Imiya, A., Iwawaki, K. and Kawamoto, K. (2002) An efficient statistical method for subpixel optical flow detection. *Engineering Applications of Artificial Intelligence*, **15** (2), 169–176.

[10] Horn, B.K.P. and Schunck, B.G. (1981) Determining optical flow. *Artificial Intelligence*, **17**, 185–203.

[11] Verri, A. and Poggio, T. (1989) Motion field and optical flow: qualitative properties. *IEEE Transactions on Pattern Analysis and Machine Intelligence*, **11** (5), 490–498.

[12] Brox, T. and Malik, J. (2011) Large displacement optical flow: descriptor matching in variational motion estimation. *IEEE Transactions on Pattern Analysis and Machine Intelligence*, **33** (3), 500–513.

[13] Fragkiadaki, K., Hu, H. and Shi, J. (2013) Pose from flow and flow from pose. Computer Vision and Pattern Recognition, 2013, pp. 2059–2066.

[14] Zuffi, S., Romero, J., Schmid, C. and Black, M.J. (2013) Estimating human pose with flowing puppets. International Conference on Computer Vision.

[15] Ferrari, V., Marin-Jimenez, V. and Zisserman, A. (2009) 2D human pose estimation in TV shows. LNCS Statistical and Geometrical Approaches to Visual Motion Analysis, pp. 128–147.

[16] Hilsmann, A. and Eisert, P. (2007) Deformable object tracking using optical flow constraints. European Conference on Visual Media Production, 2007, pp. 1–8.

[17] Brox, T., Rosenhahn, B., Cremers, D. and Seidel, H.P. (2006) High accuracy optical flow serves 3-D pose tracking: exploiting contour and flow based constraints. European Conference on Computer Vision.

[18] Fablet, R. and Black, M.J. (2002) Automatic detection and tracking of human motion with a view-based representation. European Conference on Computer Vision.

[19] Brox, T., Bruhn, A., Papenberg, N. and Weickert, J. (2004) High accuracy optical flow estimation based on a theory for warping. *European Conference on Computer Vision*, **3024**, 25–36.

[20] Bloom, V., Makris, D. and Argyriou, V. (2012) G3d: a gaming actions colour, depth and pose dataset with a real time action recognition evaluation framework. CVPR-CVCG.

[21] Yalcin, H., Hebert, M., Collins, R. and Black, M.J. (2005) A flow-based approach to vehicle detection and background mosaicking in airborne video. Computer Vision and Pattern Recognition, Vol. **2**, p. 1202.

[22] Matsushita, Y., Ofek, E., Tang, X. and Shum, H.Y. (2005) *Full-Frame Video Stabilization*, ACM.

[23] Kanatani, K., Ohta, N. and Shimizu, Y. (2002) 3D reconstruction from uncalibrated-camera optical flow and its reliability evaluation. *Systems and Computers in Japan*, **33** (9), 1665–16662.

[24] Jepson, A. and Black, M. (1993) Mixture models for optical flow computation. Computer Vision and Pattern Recognition, pp. 760–761.

[25] Weiss, Y. (1997) Smoothness in layers: motion segmentation using nonparametric mixture estimation. Computer Vision and Pattern Recognition, pp. 520–526.

[26] Talukder, A. and Matthies, L. (2004) Real-time detection of moving objects from moving vehicles using dense stereo and optical flow. *International Conference on Intelligent Robots and Systems*, **4**, 3718–3725.

[27] Torr, P. and Zisserman, A. (2000) Feature based methods for structure and motion estimation. *Vision Algorithms: Theory and Practice*, LNCS 1883, Springer-Verlag, pp. 278–294.

[28] Canny, J. (1986) A computational approach to edge detection. *IEEE Transactions on Pattern Analysis and Machine Intelligence*, **8** (6), 679–698.

[29] Marr, D. and Hildreth, E. (1980) Theory of edge detection. *Proceedings of the Royal Society of London, Series B: Biological Sciences*, **207** (1167), 215–217.

[30] Harris, C. Stephens, M. (1988) A combined corner and edge detector. Proceedings of the 4th Alvey Vision Conference, pp. 147–151.

[31] Lowe, D.G. (1999) Object recognition from local scale-invariant features. Proceedings of the International Conference on Computer Vision, Vol. **2**, pp. 1150–1157.

[32] Haar, A. (1910) Zur Theorie der orthogonalen Funktionensysteme. *Mathematische Annalen*, **69** (3), 331–371.

[33] Jawahar, C.V. and Narayanan, P.J. (2000) Feature integration and selection for pixel correspondence. Indian Conference on Computer Vision, Graphics and Image Processing.

[34] Addison, P.S. (2002) *The Illustrated Wavelet Transform Handbook*, Institute of Physics.

[35] Freeman, W. and Adelson, E. (1991) The design and use of steerable filters. *IEEE Transactions on Pattern Analysis and Machine Intelligence*, **13**, 891–906.

[36] Lazebnik, S., Schmid, C. and Ponce, J. (2004) Semi-local affine parts for object recognition. Proceedings of the British Machine Vision Conference, 2004.

[37] Kim, S., Yoon, K. and Kweon, I.S. (2006) Object recognition using a generalized robust invariant feature and gestalt's law of proximity and similarity. Conference on Computer Vision and Pattern Recognition Workshop, 2006.

[38] Ke, Y. and Sukthankar, R. (2004) PCA-SIFT: a more distinctive representation for local image descriptors. Computer Vision and Pattern Recognition, 2004.

[39] Mikolajczyk, K. and Schmid, C. (2005) A performance evaluation of local descriptors. *IEEE Transactions on Pattern Analysis and Machine Intelligence*, **27** (10), 1615–1630.

[40] Bay, H., Tuytelaars, T. and Van Gool, L. (2006) SURF: speeded up robust features. Proceedings of the 9th European Conference on Computer Vision, 2006.

[41] Oyallon, E. and Rabin, J. (2013) An analysis and implementation of the SURF method, and its comparison to SIFT, Image Processing On Line.

[42] Dalal, N. and Triggs, B. (2005) Histograms of oriented gradients for human detection. Computer Vision and Pattern Recognition, pp. 886–893.

[43] Bourdev, L. and Malik, J. (2009) Poselets: body part detectors trained using 3D human pose annotations. International Conference on Computer Vision.

[44] Wang, L. and He, D.C. (1990) Texture classification using texture spectrum. *Pattern Recognition*, **23** (8), 905–910.

[45] Leutenegger, S., Chli, S. and Siegwart, S. (2011) Brisk: binary robust invariant scalable keypoints. International Conference on Computer Vision.

[46] Alahi, A., Ortiz, R. and Vandergheynst, P. (2012) FREAK: fast retina keypoint. Computer Vision and Pattern Recognition, pp. 510–517.

[47] Rublee, E. *et al.* (2011) ORB: an efficient alternative to SIFT or SURF. International Conference on Computer Vision.

[48] Calonder, M. *et al.* (2010) Brief: binary robust independent elementary features. European Conference on Computer Vision, pp. 778–792.

[49] Schaeffer, C. (2013) A Comparison of Keypoint Descriptors in the Context of Pedestrian Detection: FREAK vs. SURF vs. BRISK.

[50] Patel, A., Kasat, D.R., Jain, S. and Thakare, V.M. (2014) Performance analysis of various feature detector and descriptor for real-time video based face tracking. *International Journal of Computer Applications*, **93** (1), 37–41.

[51] Duda, R.O. and Hart, P.E. (1972) Use of the hough transformation to detect lines and curves in pictures. *Commun. ACM*, **15**, 11–15.

[52] Tomasi, C. and Kanade, T. (1991) Detection and Tracking of Point Features. Carnegie Mellon University Technical Report, CMU-CS-91-132.

[53] Al-Najdawi, N., Tedmori, S., Edirisinghe, E. and Bez, H. (2012) An automated real-time people tracking system based on KLT features detection. *International Arab Journal of Information Technology*, **9** (1), 100–107.

[54] Keni, B. and Rainer, S. (2008) Evaluating multiple object tracking performance: the clear mot metrics. *EURASIP Journal on Image and Video Processing*, **2008**, 1–10.

[55] Han, B., Roberts, W., Wu, D. and Li, J. (2007) Robust feature-based object tracking. Proceedings of SPIE, Vol. 6568.

[56] Galvin, B., McCane, B. and Novins, K. (1999) Robust feature tracking. 5th International Conference on Digital Image Computing, Techniques, and Applications, 1999, pp. 232–236.

[57] Krinidis, M., Nikolaidis, N. and Pitas, I. (2007) 2-D feature-point selection and tracking using 3-D physics-based deformable surfaces. *IEEE Transactions on Circuits and Systems for Video Technology*, **17** (7), 876–888.

[58] Shi, J. and Tomasi, C. (1994) Good features to track. Conference on Computer Vision and Pattern Recognition, 1994, pp. 593–600.

[59] Tzimiropoulos, G. and Pantic, M. (2014) Gauss-Newton deformable part models for face alignment in-the-wild. CVPR.

[60] Li, N., Liu, L. and Xu, D. (2008) Corner feature based object tracking using Adaptive Kalman Filter. International Conference on Signal Processing, pp. 1432–1435.

[61] Xu, S. and Chang, A. (2014) Robust object tracking using Kalman filters with dynamic covariance. SPIE.

[62] Fazli, S., Pour, H.M. and Bouzari, H. (2009) Particle filter based object tracking with sift and color feature. International Conference on Machine Vision, pp. 89–93.

[63] Qi, Z., Ting, R., Husheng, F. and Jinlin, Z. (2012) Particle filter object tracking based on harris-SIFT feature matching. *Proc. Eng.*, **29**, 924–929.

[64] Schikora, M., Koch, W. and Cremers, D. (2011) Multi-object tracking via high accuracy optical flowand finite set statistics. International Conference on Acoustics, Speech and Signal Processing, Vol. **1409**, p. 1412.

[65] Benfold, B. and Reid, I. (2011) Stable multi-target tracking in real-time surveillance video. Computer Vision and Pattern Recognition.

[66] Medrano, C., Herrero, J.E., Martinez, J. and Orrite, C. (2009) Mean field approach for tracking similar objects. *Computer Vision and Image Understanding*, **113** (8), 907–920.

[67] Kuhn, H.W. (1955) The Hungarian method for the assignment problem. *Naval Research Logistic Quarterly*, **2**, 83–97.

[68] Cox, I. and Hingorani, S. (1996) An efficient implementation of reid's multiple hypothesis tracking algorithm and its evaluation for the purpose of visual tracking. *IEEE Transactions on Pattern Analysis and Machine Intelligence*, **18** (2), 138–150.

[69] McLaughlin, N., Martinez, J. and Miller, P. (2013) Online multiperson tracking with occlusion reasoning and unsupervised track motion model. IEEE International Conference on Advanced Video and Signal Based Surveillance.

[70] Language and Media Processing Laboratory. ViPER: The Video Performance Evaluation Resource, http://viper-toolkit.sourceforge.net/ (accessed 12 February 2015).

[71] CHIL. Computers in the Human Interaction Loop, http:// chil.server.de/.

[72] AMI Consortium. AMI: Augmented Multiparty Interaction, http://www.amiproject.org/ (accessed 12 February 2015).

[73] VACE. Video Analysis and Content Extraction, http://www.informedia.cs.cmu.edu/arda/vaceII.html (accessed 12 February 2015).

[74] ETISEO. Video Understanding Evaluation, http://www.silogic.fr/etiseo/.

[75] CAVIAR. CAVIAR: Context Aware Vision using Image-Based Active Recognition, http://homepages.inf.ed.ac.uk/rbf/CAVIAR/ (accessed 12 February 2015).

[76] Ziliani, F., Velastin, S., Porikli, F. *et al.* (2005) Performance evaluation of event detection solutions: the CREDS experience. Proceedings of the IEEE International Conference on Advanced Video and Signal Based Surveillance, 2005, pp. 201–206.

[77] PETS. PETS: Performance Evaluation of Tracking and Surveillance, http://www.cbsr.ia.ac.cn/conferences/VS-PETS-2005/

[78] Sigal, L., Balan, A. and Black, M.J. (2010) HumanEva: synchronized video and motion capture dataset and baseline algorithm for evaluation of articulated human motion. *International Journal of Computer Vision*, **87** (1-2), 4–27.

[79] EEMCV. EEMCV: Empirical Evaluation Methods in Computer Vision, http://www.cs.colostate.edu/eemcv2005/.

[80] AVSS. Advanced Video and Signal-Based Surveillance, http://www.avss2013.org/ (accessed 12 February 2015).

[81] CLEAR. Classification of Events, Activities and Relationships, http://www.clear-evaluation.org/

[82] The i-LIDS Dataset, http://scienceandresearch.homeoffice.gov.uk/hosdb/cctv-imaging-technology/video-based-detection-systems/i-lids/.

[83] Burr, J., Mowatt, G., Siddiqui, M.A.R., Herandez, R. *et al.* (2007) The clinical and cost effectiveness of screening for open angle glaucoma: a systematic review and economic evaluation. *Health Technology Assessment*, **11** (41), 1–190.

[84] Guerreiro, R. and Aguilar, P. (2006) Global motion estimation: feature-based, featureless, or both ?! *Image Analysis and Recognition*, LNCS 4141, Springer-Verlag, pp. 721–730.

[85] Shi, J. and Tomasi, C. (1994) Good features to track. Computer Vision and Pattern Recognition, 1994, pp. 593–600.

[86] Sinha, S.N., Frahm, J.M., Pollefeys, M. and Gen, Y. (2007) Feature tracking and matching in video using programmable graphics hardware. Machine Vision and Applications.

4

Face Alignment and Recognition Using Registration

4.1 Introduction

Face recognition is one of the fastest growing and most consolidated biometric identification systems in the last decades. It is currently used for security and surveillance applications, such as airport and border control, counter-terrorism measurements, visa application processing, access control or police CCTV surveillance, and it is also applied to many other diverse fields, such as automatic tagging on social networks, automatic focus on digital photography or real-time feeling detection. This has been possible, thanks to the ubiquity of photo and video cameras, the non-intrusive characteristics of face recognition and the increase in performance of computer vision and machine-learning algorithms.

Among the different steps involved in face recognition, face alignment is a crucial step in order to deal with real environments such as the previously mentioned. While many recognition algorithms perform well on public datasets, they degrade drastically in real application due to changes in pose, shape or illumination. Given the increasingly exigence of current application, with vast number of users to be recognised, and the consequent accuracy required, the decrease in performance caused by misalignment cannot be afforded, making face alignment a crucial part of the pipeline. In fact, it has been stated that an alignment error cannot be rectified in later steps [1]. A clear example of how face alignment can affect the performance of a recognition algorithm is depicted in Figure 4.1. The necessity of a robust alignment algorithm is even more evident in those application where the subject is not cooperative or it is not aware of being monitored, such as security ones. Since one of the selling points of face recognition is precisely its potential to work in less control scenarios, alignment reveals as a compulsory element.

Face alignment is a non-rigid registration problem, where an unseen face is intended to be matched against one or several canonical models or templates. In comparison

Image, Video & 3D Data Registration: Medical, Satellite & Video Processing Applications with Quality Metrics,
First Edition. Vasileios Argyriou, Jesus Martinez Del Rincon, Barbara Villarini and Alexis Roche.
© 2015 John Wiley & Sons, Ltd. Published 2015 by John Wiley & Sons, Ltd.

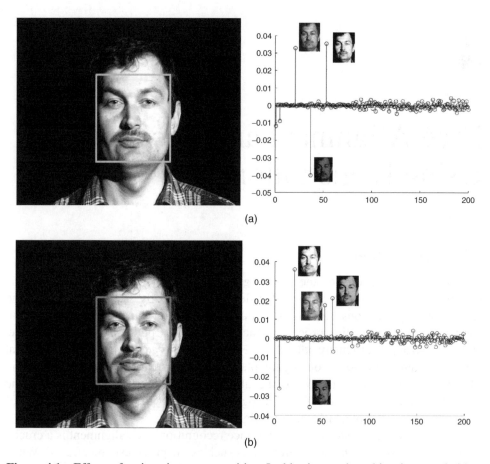

(a)

(b)

Figure 4.1 Effects of registration on recognition. In this picture, the subject is wrongly identified when using directly the input provided by Viola and Jones face detector (box in a). In (b) the person is correctly identified after correcting the alignment and compensating the illumination

with other registration problems, it entails a high degree of complexity due to its non-rigid nature and the large variability of faces in shapes, expressions, facial accessories or lighting conditions.

4.2   Unsupervised Alignment Methods

Unsupervised registration methods comprise all those methodologies where no prior information, manual labelling or high-level understanding of the facial image are assumed or considered beforehand. While this is indeed a clear advantage, a cost is paid in terms of computational performance, since the number of points involved in the registration is at least an order of magnitude higher than other methodologies. They are also more prone to errors due to large occlusions, such as wearing scarfs or sunglasses, since those situations are more difficult to identify.

Although a few initial approaches have tried to use raw pixels both in colour or grey level as unsupervised features, they have been rapidly overtaken but in other methods where more salient features were previously extracted. Unsupervised methods include global feature methods or automatically defined local features (key points).

4.2.1 Natural Features: Gradient Features

The use of grey-level edge maps allows capturing the structure of salient image features better by addressing the low-pass nature of facial images, interpolation errors, border effects and aliasing. In [2], a normalized gradient correlation (NGC), is used to estimate the unknown parameters of an affine transformation in the log-polar Fourier domain. The usage of NGC allows rejecting outliers induced in the face registration world by real-world environments:

$$
\text{NGC}(u) \triangleq q \frac{G_1(u) \star G_2^*(-u)}{|G_1(u)| \star |G_2(-u)|} = \frac{\int_{R^2} G_1(x) G_2^*(x+u) dx}{\int_{R^2} |G_1(x) G_2(x+u)| dx} \tag{4.1}
$$

where

$$
G_i(x) = G_{i,x}(x) + j G_{i,y}(x) \tag{4.2}
$$

being $G_{i,x} = \nabla_x I_i$ and $G_{i,y} = \nabla_y I_i$ the gradients of a given image along the horizontal and vertical directions, respectively.

A more detailed explanation on frequency domain registration methods in general and how this normalized correlation is integrated in the registration process are depicted in Chapter 5.

4.2.2 Dense Grids: Non-rigid Non-affine Transformations

As we explained in previous chapters, the classical registration approach allows us to state the registration problem as a energy minimization problem so that

$$
E(\Delta \mathbf{x}) = \sum_{\mathbf{x}} (I_1(W(\mathbf{x}, \Delta \mathbf{x})) - I_T(\mathbf{x}))^2 \tag{4.3}
$$

where $\mathbf{x} = (x, y)$ are the pixel coordinates, $\Delta \mathbf{x}$ are the registration parameters, I_1 is the input image and I_T the template reference image, and $W(\mathbf{x}, \Delta \mathbf{x})$ is some parametric warping function between the images at location $\mathbf{x}$.

However, an affine-warping function W, although sufficient for many other rigid registration problems, will not perform well in face alignment because of its multiple variability sources. Recent examples of these approaches applied to faces can be found in [3, 4], which treat the entire face at once under the assumption that an affine transformation is enough to solve the problem. The consequence is a limited capacity to deal with variability, which is reflected on dealing uniquely with frontal faces or with neutral expressions. As a solution to mitigate the problem, faces can be divided into patches that are processed independently [5]. Although this strategy improves the performance under more severe sources of variability such as poses, it struggles to keep consistency over neighbour patches resulting in poor looking and non photo-realistic results.

An alternative solution is presented in [6]. By replacing the warping function by a parameterized transformation of a 2D mesh vertex coordinate, an unsupervised face alignment, capable of robustly dealing with non-rigid variations, is designed. Instead of registering the displacement of every point on the face, only those keypoints belonging to the mesh will be registered. Under this framework, a triangular mesh s is defined by its vertex, so

$$s = [x_1, x_2, \ldots, x_N, y_1 y_2, \ldots, y_N]^\top \tag{4.4}$$

This notation is introduced into Equation (4.3) by replacing the estimation of the registration parameters Δx with the estimation of the new location of the vertex Δs

$$E(\Delta s) = \sum_x (I_T(x, \Delta s) - I_1(W(x, s)))^2 \tag{4.5}$$

As usual in registration, this problem is ill-conditioned due to the need of estimating $2N$ variables. In order to solve the consequent aperture problem, a constraint [7] is introduced for regularization purposes. Therefore, the final equation is stated as follows:

$$E(\Delta s) = \sum_x (I_T(x, \Delta s) - I_1(W(x, s)))^2 + \lambda_s s^\top K s \tag{4.6}$$

where Δs is the increment to the mesh vertices and K is a sparse $2N \times 2N$ matrix containing the connectivity of the mesh.

By applying first-order Taylor expansion, the previous equation can be linearized such as

$$E(\Delta s) \approx \sum_x (I_T + A\Delta s - I_1(s)))^2 + \lambda_s (s + \Delta s)^\top K(s + \Delta s) \tag{4.7}$$

where $A = \nabla I_T \frac{\partial W}{\partial s}$ is the steepest descent image, being ∇I_T the gradient of the reference image and $\frac{\partial W}{\partial s}$ the Jacobian of the warping parameters.

Thus, the estimated alignment of the mesh Δs can be computed through optimization so

$$\Delta s = H^{-1} \left(\sum_x A^\top (I_1(s) - I_T) - \lambda_s K s \right) \tag{4.8}$$

where H is the Hessian matrix

$$H = \sum_x A^\top A + \lambda_s K \tag{4.9}$$

Finally, in order to improve the performance of the previous approximation against intensity outliers, a robust estimator $\rho(u)$ can be added to the previous methodology. In this way, the energy minimization problem to be solved can be written as

$$E(s) = \sum_x \rho(I_T(x, \Delta s) - I_1(W(x, s)))^2 + \lambda_s s^\top K s \tag{4.10}$$

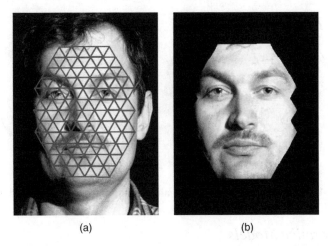

(a) (b)

Figure 4.2 Dense grid registration. (a) Reference image with grid overlaying and extracted reference template. (b) Input image with warped resulting grid and registered input image [6]

giving the solution

$$\Delta s = H_R^{-1} \left(\sum_{\mathbf{x}} \rho\prime(u^2) A^\top (I_1(\mathbf{s}) - I_T) - \lambda_s K \mathbf{s} \right) \tag{4.11}$$

where $\rho\prime$ is the first-order derivative of the robust estimator and H_R is the Hessian matrix

$$H_R = \sum_{\mathbf{x}} \rho\prime(u^2) A^\top A + \lambda_s K \tag{4.12}$$

Many examples of kernels or robust estimators can be applied and be found in the literature, depending on our particular noise, such as exponential functions or loss functions (Figure 4.2).

4.3 Supervised Alignment Methods

Supervised alignment methods encapsulate those registration methods that use a set of key points or landmarks that have been manually labelled. By using this manual labelling, we intend to facilitate the extraction of consistent feature points not only for the alignment but also for the posterior recognition tasks. This supervision ensures high registration quality given the high variance and non-rigid nature of the face, where consistent points are difficult to localize. While facial images contain less reliable and persistent points than other rigid objects, a few of them, such as eye and mouth corners, can still be consistently detected by an automatic method (Figure 4.3).

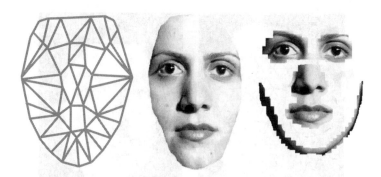

Figure 4.3   Examples of supervised alignment methods for face recognition

4.3.1   Generative Models

Generative methods are those face registration techniques that make use of a statistical shape model built from representative training samples. The match of a model with unseen data reduces the number of points to use, filters false detections and ensures the consistency among the keypoints according to maximum variability learned during training.

The active appearance models (AAMs) [8] and active shape models (ASMs) [9] are probably the most widely used generative techniques in face recognition applications. AAMs match parametric models of shape and appearance into unseen input images by solving a nonlinear optimization that minimizes the difference between the synthetic reference template and the real appearance. These models not only provide high-quality registration points but also represent relevant facial information with a small set of model parameters. By carefully selecting the training images, they are able to learn most of the variability sources such as change of expressions, poses and/or illumination.

In these approaches, the non-rigid shape of the face is modelled as a linear combination of a rigid mean shape vector and a set of n basis shapes. This formulation is also called point distribution model (PDM) [9]. The shape itself is defined by a set of points or vertexes, which confirm a mesh $\mathbf{s} = [s_i]' \forall i \in [1, N]$. These keypoints are manually selected to be important structural facial features, such as eye corners, eyebrows, mouth, nose, chin, to name a few (see Figure 4.4(a)), and the shape model is learnt from a representative dataset manually annotated.

In order to learn the shape model $\mathbf{s}$, the shapes are aligned into a common mean shape using a generalized procrustes analysis (GPA) [10] that removes location, scale and rotation effects (see Figure 4.4(b)). Applying a Principal Components Analysis (PCA) to the aligned shapes, the parametric model is learned so $\mathbf{s} = \mathbf{s}_0 + \phi p$, where p is the vector of the shape configuration weights, $\mathbf{s}_0$ is the mean shape and the ϕ is a matrix containing the relevant eigenvectors, which represent the allowed models of deformation. Since PCA can be understood a dimensionality reduction technique, p is also referred as the low-dimensionality representation of the shape in the trained

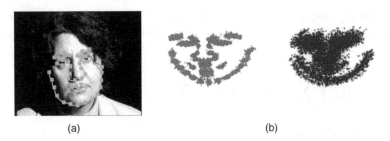

(a)   (b)

Figure 4.4 (a) Typical keypoints used in generative models. (b) Training set samples used for learning a PDM model before and after applying procrustes algorithm [9]

space. This model has been extended in other works to include an extra parameter that allows modelling more subtle changes in pose and expression [11].

The appearance model A is learned by first warping each sample to the mean shape using a warping function $W(\mathbf{x}, p)$, being $\mathbf{x}$ the facial pixels. This warp function is a piecewise affine warp that is function of the shape. Each piece corresponds, therefore, to a triangle of the mesh, which is calculated using Delaunay triangulation. The warped image $I_1(W(\mathbf{x}, p))$ is computed by backwards warping the input image $I_1(\mathbf{x})$. Afterwards, the appearance model A is calculated by applying PCA on all warped images, so

$$A(\mathbf{x}) = A_0(\mathbf{x}) + \sum_{i=1}^{m} \lambda_i A_i(\mathbf{x}) \qquad (4.13)$$

where A_0 is the mean appearance, A_i is the eigen image and λ is the appearance parameter.

4.3.1.1 Fitting the Model

Fitting the model to a new unseen input image is itself the registration problem to be solved. Initially, non-gradient descent algorithms were applied for dealing efficiently with piecewise affine warps [8], but more efficient gradient descent techniques have been recently developed.

As a registration problem, the goal is to minimize the previously stated Equation (4.3). Given that inverse compositional approaches (IC) [12] are invalid for piecewise warping [13], a similar reasoning in Section 2.2 can be followed using the additive formulation proposed by Lucas–Kanade.

The simultaneous forwards additive (SFA) [11] algorithm searches for shape and appearance parameters simultaneously by minimizing the squared differences between the appearance model and the input-warped image. The optimization consists in solving

$$\arg \min_{p, \lambda} \sum_{\mathbf{x}} [A_0(\mathbf{x}) + \sum_{i=1}^{m} \lambda_i A_i(\mathbf{x}) - I_1(W(\mathbf{x}, p))]^2 \qquad (4.14)$$

By expanding the previous equation using first-order Taylor expansion and using gradient descent additives update, the solution for the combined parameters $r = [p^T, \lambda^T]^T$ update Δr is

$$\Delta r = H_{sfa}^{-1} \left(\sum_{\mathbf{x}} SD(\mathbf{x})_{sfa}^T (A_0(\mathbf{x}) + \sum_{i=1}^{m} \lambda_i A_i(\mathbf{x}) - I_1(W(\mathbf{x}, p))) \right) \qquad (4.15)$$

where H_{sfa} is the Hessian matrix

$$H_{sfa} = \sum_{\mathbf{x}} SD(\mathbf{x})_{sfa}^T SD(\mathbf{x})_{sfa} \qquad (4.16)$$

and $SD(\mathbf{x})$ is the Steepest Descent image:

$$SD(\mathbf{x})_{sfa} = \left[\nabla I_1 \frac{\partial W}{\partial p_1} \cdots \nabla I_1 \frac{\partial W}{\partial p_n} - A_1(\mathbf{x}) \cdots - A_m(\mathbf{x}) \right] \qquad (4.17)$$

The SFA algorithm is slow but very accurate since it searches simultaneously for shape and appearance parameters, which may require a large number of iterations. In order to obtain a faster but less accurate fitting, the normalization forwards additive (NFA) [11] can be applied instead. The idea behind consists in assuming initially a fixed appearance so that it searches only for the shape parameters. The appearance is projected from the residual error.

Under a fix appearance assumption, $A(\mathbf{x}, \lambda) = A_0(\mathbf{x})$ and $SD(\mathbf{x})$ can be written as

$$SD(\mathbf{x})_{nfa} = \left[\nabla I_1 \frac{\partial W}{\partial p_1} \cdots \nabla I_1 \frac{\partial W}{\partial p_n} \right] \qquad (4.18)$$

resulting in a simplified H_{nfa} Hessian matrix

$$H_{nfa} = \sum_{\mathbf{x}} SD(\mathbf{x})_{nfa}^T SD(\mathbf{x})_{nfa} \qquad (4.19)$$

The appearance parameters λ can be extracted in isolation as

$$\lambda_i = \sum_{\mathbf{x}} A_i(\mathbf{x})(A_0(\mathbf{x}) - I_1(W(\mathbf{x}, p)))) \qquad (4.20)$$

while the final solution for the remaining parameters is

$$\Delta p = H_{nfa}^{-1} \left(\sum_{\mathbf{x}} SD(\mathbf{x})_{nfa}^T (A_0(\mathbf{x}) - I_1(W(\mathbf{x}, p))) - \sum_{i=1}^{m} \lambda_i A_i(\mathbf{x})) \right) \qquad (4.21)$$

4.3.2 Discriminative Approaches

The discriminative approaches follow a bottom–up approach in contrast to the top–down approach of generative models. These techniques use an ensemble of feature detectors that aim to search for each facial landmark.

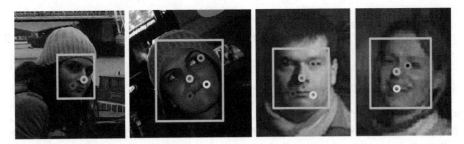

Figure 4.5 Keypoints corresponding to eye, mouth corners and nose are extracted using dedicated classifiers and a filtering strategy to remove false positives

An example of this pipeline is explained in [14]. A set of key points are automatically extracted using a Viola and Jones' style boosted classifier with Haar-line image features. Five classifiers are applied to detect candidates for the left and right eyes, the nose tip and two mouth corners. Since discriminative approaches are likely to obtain multiple candidates for each keypoint, a filtering strategy is needed to remove the less consistent points with a hypothetical face. Examples of the extracted keypoint candidates are shown in Figure 4.5.

The filtering strategy first removes possible false positives by applying heuristic rules related with face symmetry and morphology. In a second step, a simple probabilistic Gaussian model evaluates the candidate points regarding their spatial 2D distribution. A different and also conventional alternative in the literature consists of keeping this consistency by selecting the local candidates connected by a shape regularization model.

Finally, once a set of consistent keypoints are available and given their reduce number, the affine transformation of the face can be calculated by linear least squares.

Let $\mathbf{x}_i = [x_i^1, y_i^1]$ and $\mathbf{y}_i = [x_i^T, y_i^T]$ be the coordinates of a keypoint $i \in [1, 5]$ in the input image I_1 and template reference image I_T, respectively. A similarity transformation can be represented as

$$\mathbf{y}_i = R * \mathbf{x}_i + t \tag{4.22}$$

where $t = [t_x t_y]^\top$ is the translation vector and $R = [r_1, -r_2; r_2, r_1]$ contains the scale and rotation parameters.

Under this formulation, the transformation parameters can be extracted by solving an overdetermined system of linear equations $A_p = B$:

$$
\begin{bmatrix}
x_1^1 & -y_1^1 & 1 & 0 \\
y_1^1 & x_1^1 & 0 & 1 \\
\vdots & \vdots & \vdots & \vdots \\
x_i^1 & -y_i^1 & 1 & 0 \\
y_i^1 & x_i^1 & 0 & 1 \\
\vdots & \vdots & \vdots & \vdots
\end{bmatrix}
\begin{bmatrix}
r_1 \\
r_2 \\
t_x \\
t_y
\end{bmatrix}
=
\begin{bmatrix}
x_1^T \\
y_1^T \\
\vdots \\
x_i^T \\
y_i^T \\
\vdots
\end{bmatrix}
\tag{4.23}
$$

using least squares:

$$p = (A^T A)^{-1} A^T B \qquad (4.24)$$

4.4 3D Alignment

The increasing accuracy requirements for real-life application, the appearance and expansion of cheap 3D sensors and stereo cameras, and the necessity of researchers for differentiating have promoted the appearance of 3D face recognition techniques and consequently 3D face alignment algorithms. On the one hand, they have greater potential than 2D techniques since they consider more information simultaneously - such as depth and curvature - they are naturally pose invariant and they solve the self-occlusion problem. On the other hand, the alignment is computationally more intensive and normally requires a good initialization.

4.4.1 Hausdorff Distance Matching

Hausdorff distance is a well-known technique in computer vision that allows us to match two objects under occlusions. Formally, it measures how far two sets of points are from each other, since the two sets are closed only if every point in one of the sets is close to some point in the other set.

Hausdorff distance has been commonly used in 3D face alignment [15, 16] as a registration metric to match the acquired 3D points of a face against another 3D point reference face. Let $I_1^{3D} = [x_1, \ldots, x_N]$ and $I_T^{3D} = [y_1, \ldots, y_N]$ be two sets of 3D points such as $x_i, y_i \in \mathcal{R}^3$. The Hausdorff distance is defined as

$$d_H(I_1^{3D}, I_T^{3D}) = \max(d_h(I_1^{3D}, I_T^{3D}), d_h(I_T^{3D}, I_1^{3D})) \qquad (4.25)$$

where d_h is the directed Hausdorff distance

$$d_h(I_1, I_2) = \max_{x \in I_1} \min_{y \in I_2} \| x - y \| \qquad (4.26)$$

Intuitively, this metric can be understood as the greatest of all distances from a point in one of the 3D images to the closest point in the reference 3D image.

As additional advantage, the Hausdorff distance can be extended to handle occluded points by modifying the directed Hausdorff distance:

$$d_h(I_1, I_2) = K^{th}_{x \in I_1} \min_{y \in I_2} \| x - y \| \qquad (4.27)$$

where $K^{th}_{x \in I_1}$ is the Kth ranked value in the set of distances, and K controls the maximum number of points to be occluded.

4.4.2   Iterative Closest Point (ICP)

The ICP algorithm solves the registration problem by iteratively minimizing the distance between a reference set of points I_T^{3D} and an input set I_1^{3D}, which is transformed to best match the reference.

The algorithm steps are as follows:

- For each point in the input 3D set $\forall \mathbf{x}_i \in I_1^{3D}$, find the closet point $\mathbf{y}_j$ in the reference set $\arg \min_{j \in N} \| \mathbf{x}_i - \mathbf{y}_j \|$.
- Estimate the rigid transformation (R,t) that best aligns each input point from the previous step using the mean square error (MSE)

$$\text{MSE}(R, t) = \frac{1}{N} \sum_{i=1}^{N} \| (R * \mathbf{x}_i + t) - \mathbf{y}_j \| \qquad (4.28)$$

- Transform the input points using the obtained transformation $\mathbf{x}\prime_i = R * \mathbf{x}_i + t$.
- Reiterate.

The algorithm ensures convergence to a local minimum monotonically, but its convergence to a global minimum depends on the quality of the initialization. To ensure that no residual error is computed from the non-overlapping areas of the point sets, those pairs containing points on surface boundaries are usually rejected.

4.4.3   Multistage Alignment

While ICP has been extensively used for 3D face alignment [17, 18], it is ill-suited for the problem, given the required good initialization and the non-rigid nature of the face and expression changes, which clash with the estimated rigid transformation. To address this issue, many variations and multistage alignment algorithms comprising ICP have been proposed [19, 20] as well as alignments on certain regions of the face less prone to expression changes.

In [1], a three-stage alignment algorithm is proposed, targeting the achievement of a high degree of expression invariance. The alignment is therefore going from coarse to fine, every stage offering less resilience to local minimum while increasing the accuracy.

- *SpinImages [21]*: First, a course initialization is calculated. For each point belonging to both point sets, a spin image representing its neighbourhood is computed, and the consistent correspondences between spin images are extracted.
- *ICP:* ICP algorithm is applied to the output of the previous stage.
- *Simulated annealing on z-buffers:* The alignment is refined by minimizing the differences between both reference and input Z-buffers. This is achieved by applying a global optimization methodology called Enhanced Simulated Annealing [22].

4.5   Metrics for Evaluation

4.5.1   Evaluating Face Recognition

Due to the lack of specific and annotated datasets, the classical face alignment evaluation has been performed indirectly, in conjunction with the face recognition tasks. By obtaining a final recognition rate and plugging/unplugging the registration step, the effectiveness of the alignment could be measured.

Under this evaluation framework, two different scenarios can be used to measure the performance in recognition. In an identification scenario, the goal is to identify a person within a database. The performance of the system can be measured using the accuracy rate, which can be simplified, since no negative response is expected from the classifier (an answer will be always given).

$$\text{Accuracy} = \frac{\text{TP} + \text{TN}}{\text{TP} + \text{FN} + \text{FP} + \text{TN}} = \frac{\text{TP}}{\text{TP} + \text{FP}} \tag{4.29}$$

True positives TP refer to those that correctly identify subjects and false positives FP to errors.

For those application where the number of subjects is very big, and therefore the expected accuracy is rather low, it is common to measure performance using a cumulative match characteristic (CMC) curve [23]. This curve allows estimating the performance of the system when several answers are affordable to be later verified. In this curve, an identity is considered a hit if it is among the first n-most likely answers of the system (rank-n). The rank-1 recognition rate of the CMC is equivalent to the accuracy.

In a verification scenario, the goal is, given a second face image, to decide whether these faces belong to the same person or not. Since the result of the comparison must be evaluated against a threshold, the trade-off between positive hits and false alarms can be easily moved by changing the threshold value. In order to avoid overestimating the performance of the system, a complete characterization can be achieved by plotting the receiver operating characteristics (ROC) curve [24]. This curve plots TP (on the y-axis) against FP (on the x-axis). The performance for any given threshold is therefore a point in the curve.

4.5.2   Evaluating Face Alignment

The previous methods are not ideal since they do not measure the performance of the registration directly but through indirect metrics. This means that the real performance can be shadowed by the performance of the classifier, especially if this one is poor.

The RMS error is the most direct way to evaluate the alignment error between the input face after registration $I\prime_1$ and the groundtruth face.

$$\text{RMS}(I_1'(\mathbf{x}), I_T(\mathbf{y})) = \sqrt{\frac{\sum_{i=1}^{N} (\mathbf{x}_i - \mathbf{y}_i)^2}{N}} \qquad (4.30)$$

This metric requires an annotated database, where each image has been carefully annotated (manually through external sensors such as MoCap). It is easy to apply for methodologies based on keypoints, where the number of points to annotate and compute is small, but almost infeasible for unsupervised methodologies. The usage of 3D sensors during the database acquisition, able to reconstruct the face in 3D, makes possible to extend the RMS to unsupervised techniques, however. These sensors also allow other metrics usage such as the deviation of the normal or the distance between two surfaces.

Recently, together with new and better annotated datasets, direct metrics have started to appear and become standard to facilitate comparison. Particularly useful is the Takeo experiment [13, 25]. In this experiment, a region of interest within the face and three canonical points in this region are selected. These points are perturbed using Gaussian noise of a given standard deviation σ, and the initial RMS error between the canonical and perturbed points is computed. An affine distorted image is generated using the affine warp that the original and perturbed points define. Given the warp estimate, the destination of the three canonical points and, similarly, the final RMS error between the estimated and correct locations can be computed.

Under this evaluation framework, the average rate of convergence for a given σ and the average frequency of convergence for a range of σ are used as the performance evaluation measures. In addition, an algorithm is considered to have converged if the final RMS point error was less than a threshold of pixels after a set number of iterations.

4.5.3 Testing Protocols and Benchmarks

The explosion of commercial products and applications using face recognition technologies has required its regulation and standardization through governmental independent evaluations and normatives.

Face Image ISO Compliance Verification Benchmark Area (FVC-onGoing) [26] is an automatic web-based evaluation system developed to evaluate face recognition algorithms. Algorithms submitted to these benchmarks are required to check the compliance of face images to ISO/IEC 19794-5 standard. It is composed of two benchmarks:

- *FICV-TEST:* A simple dataset composed of 720 images useful to test algorithm compliancy with the testing protocol (results obtained on this benchmark are only visible in the participant private area and cannot be published).

- *FICV-1.0:* A large dataset of 4868 high-resolution face images related to requirements such as blurred, bad lighting, shadows, distracters and occlusions.

The algorithms are submitted as a program file and the results are returned to the user without having access to the main body of the images or annotations.

Face recognition vendor test (FRVT) [27] (previously called face recognition technology (FERET) evaluations) is a set of large-scale independent evaluations promoted by NIST to measure progress of prototypes and commercial products on face recognition and identify future research directions. These tests were released as part as the face recognition grand challenge (FRGC).

- FRVT 2013 aimed to evaluate tasks such as One-to-many identification in a set of mugshot images, One-to-one verification of visa images, Multisample still facial imagery, Recognition of persons in video sequences, Twins, 90 ° profile view, Gender estimation (M–F), Age estimation (in years), Pose estimation and expression neutrality.
- MBE 2010 aimed to evaluate the performance on face recognition and other multimodel biometric tasks such as Face recognition on still frontal, Real-world-like high- and low-resolution imagery, Iris recognition from video sequences and off-angle images, Fusion of face and iris (at score and image levels), Unconstrained face recognition from still & video, Recognition from near infrared (NIR) & high-definition (HD) video streams taken through portals, Unconstrained face recognition from still & video.
- FRVT 2006 evaluated performance on high-resolution still images, 3D facial scans, facial videos, multisample still face images and pre-processing algorithms for alignment and illumination compensation. Reliable assessment is achieved by measuring performance on sequestered data.
- FRVT 2002 consisted of two tests: the High-computational intensity (HCInt) test and the Medium-computational intensity (MCInt) test. Both tests required the systems to be fully automatic, and manual intervention was not allowed.
- FRVT 2000 consisted of two components: the Recognition performance test and the Product usability test.
- FERET (1994, 1995, 1996) As crucial part of the FERET evaluations, the FERET database was collected in 15 sessions. It contains 1564 sets of images for a total of 14,126 images that include 1199 individuals and 365 duplicate sets of images. A duplicate set of images of the subjects was recorded leaving up to 2 years between their first and last sittings.

4.5.4   Datasets

The previous benchmarks allow us to test face recognition techniques in the wild confidently. However, there are also a variety of other datasets available, which are still being used for testing and, more importantly, for training prototypes and algorithms.

The AR databases [28] contain 126 people and 4000 images, recorded in two sessions over 2 days. Images feature frontal view faces with different facial expressions, illumination conditions and occlusions (sun glasses and scarf). The pictures were taken under strictly controlled conditions. Groundtruth contains four different manual annotations of the shape of each of the facial components (eyebrows, eyes, nose, mouth and chin) of the faces in the AR database.

Yale face database [29] contains, respectively, 165 greyscale images from 15 subjects, with different facial expressions and configurations: different facial expression or configuration: centre-light, w/glasses, happy, left-light, w/no glasses, normal, right-light, sad, sleepy, surprised and winks. The Yale Face Database B [30] contains 5760 images of 10 people under 675 viewing conditions (9 poses 64 illumination conditions), which allows to reconstruct 3D surfaces of the faces.

Multi-PIE Database, CMU, [31] contains images of 337 subjects, captured under 15 view points and 19 illumination conditions in four recording sessions for a total of more than 750,000 images. This is an extension of an original PIE database, containing 41,368 images of 68 people, each person under 13 different poses, 43 different illumination conditions and with 4 different expressions.

Surveillance Cameras Face Database [32] (SCface) images were taken in uncontrolled indoor environment using five video surveillance cameras of various qualities to mimic real-world conditions. The database contains 4160 static images (in visible and infrared spectrum) of 130 subjects. The groundtruth also contains 21 facial landmarks annotated manually by a human operator. Similarly, the databases on Face In Action (FIA) Face Video Database [33] and the McGill Real-World Face Video Database [34] also target to mimic the real-world complexity.

To increase the complexity, The Siblings Database [35] contains two datasets (high and low qualities) with images of individuals related by sibling relationships. A total of 184 and 196 individuals were recorded (92 and 98 pairs of siblings). Groundtruth provides 76 landmarks on frontal images and 12 landmarks on profile images, plus information on sex, birth date, age and the vote of a panel of human aspect regarding their sibling similarity.

YouTube Faces Database [36] contains 3 425 videos of 1 595 different people, all of them downloaded from YouTube. This dataset targets video pair-matching techniques rather than still face recognition.

Many dataset containing 3D scans are also available. PhotoFace: Face recognition using photometric stereo [37] is one of the largest 3D databases currently available. It contains 3187 sessions of 453 subjects with four different light coloured photographs, captured in two recording periods separated by 6 months. Recording was performed simulating real-world and unconstrained capture. Groundtruth comprises 11 facial landmarks manually located for alignment purposes plus metadata such as gender, facial hair, pose or expression.

The 3D Mask Attack Database (3DMAD) [38] is a face spoofing database containing 15 videos of 300 frames of 17 persons with frontal-view and neutral expression(in three different sessions under controlled conditions), recorded using Kinect for both real-access and spoofing attacks. Groundtruth contains manually annotated eye positions. Another database using Kinect as sensor is the EURECOM Kinect Face Dataset (EURECOM KFD) [39].

An excellent and complete description and those and more facial databases can be found at [40].

4.6 Conclusion

In this chapter, an overview of supervised and supervised techniques for face alignment using registration techniques is presented. Methods include the use of global features such as raw intensity pixels or gradients, as well as local features such as keypoints. Algorithms for aligning 3D face scans are also discussed. Finally, a summary of the most common databases and benchmarks is given, emphasizing the most accepted metric to evaluate face recognition and face alignment.

4.7 Exercise

This exercise is focused on face alignment using supervised methods. Using the training set, first train a PDM facial model and then apply it to the still face images. Once you have done it to the reference and the input image pair, apply a registration technique to align the two faces.

References

[1] Kakadiaris, I.A., Passalis, G., Toderici, G. *et al.* (2007) Three-dimensional face recognition in the presence of facial expressions: an annotated deformable model approach. *IEEE Transactions on Pattern Analysis and Machine Intelligence*, **29** (4), 640–649.

[2] Tzimiropoulos, G., Argyriou, V., Zafeiriou, S. and Stathaki, T. (2010) Robust FFT-based scale-invariant image registration with image gradients. *IEEE Transactions on Pattern Analysis and Machine Learning*, **10** (32), 1899–1906.

[3] Cox, M., Lucey, S., Sridharan, S. and Cohn, J. (2008) Least squares congealing for unsupervised alignment of images. *Computer Vision and Pattern Recognition*.

[4] Huang, G., Jain, V. and Miller, E. (2007) Unsupervised joint alignment of complex images. *International Conference on Computer Vision*.

[5] Ashraf, A.B., Lucey, S. and Chen, T. (2008) Learning patch correspondences for improved viewpoint invariant face recognition. *Computer Vision and Pattern Recognition*.

[6] Zhu, J., Van Gool, L. and Hoi, S. (2009) Unsupervised face alignment by robust nonrigid mapping. *Computer Vision and Pattern Recognition*.

[7] Zhu, J., Lyu, M.R. and Huang, T.S. (2009) A fast 2d shape recovery approach by fusing features and appearance. *IEEE Transactions on Pattern Analysis and Machine Intelligence*, **31** (7), 1210–1224.

[8] Cootes, T.F., Edwards, G.J. and Taylor, C.J. (2001) Active appearance models. *IEEE Transactions on Pattern Analysis and Machine Intelligence*, **23** (6), 681–685s.

[9] Cootes, T.F. and Taylor, C.J. (1992) Active shape models - smart snakes. *British Machine Vision Conference*.

[10] Goodall, C. (1991) Procrustes methods in the statistical analysis of shape. *Journal of the Royal Statistical Society, Series B*, **53** (2), 285–339.

[11] Dias Martins, P.A. (2012) Parametric face alignment: generative and discriminative approaches. PhD thesis. University of Coimbra.

[12] Baker, S. and Matthews, I. (2001) Equivalence and efficiency of image alignment algorithms. *Computer Vision and Pattern Recognition*, **1**, 1090–1097.

[13] Baker, S., Gross, R. and Matthews, I. (2003) Lucas-Kanade 20 Years on: Part 3. Tech. Rep. CMU-RI-TR-03-35, Robotics Institute Carnegie Mellon University, pp. 1–51.

[14] Hasan, M.K. and Pal, C.J. (2011) Improving alignment of faces for recognition. *EEE International Symposium on Robotic and Sensors Environments (ROSE)*, pp. 249–254.

[15] Achermann, B., Jiang, X. and Bunke, H. (1997) Face recognition using range images. *International Conference on Virtual Systems and MultiMedia*, pp. 129–136.

[16] Russ, T.D., Koch, K.W. and Little, C.Q. (2004) 3D facial recognition: a quantitative analysis. *45th Annual Meeting of the Institute of Nuclear Materials Management (INMM)*.

[17] Medioni, G. and Waupotitsch, R. (2003) Face recognition and modeling in 3D. *IEEE International Workshop on Analysis and Modeling of Faces and Gestures (AMFG 2003)*, pp. 232–233.

[18] Ben Amor, B., Ouji, K., Ardabilian, M. and Chen, L. (2005) 3D face recognition by ICP-based shape matching. *The 2nd International Conference on Machine Intelligence (ACIDCA-ICMI'2005)*.

[19] Ju, Q. and Wang, B. (2012) Hierarchical 3D face alignment based on region segmentation. *Journal of Convergence Information Technology(JCIT)*, **7** (10), 231–238.

[20] Bowyer, K.W., Chang, K. and Flynn, P. (2006) A survey of approaches and challenges in 3D and multi-modal 3D + 2D face recognition. *Computer Vision and Image Understanding*, **101**, 1–15.

[21] Johnson, A. (1997) Spin-images: a representation for 3-D surface matching. PhD dissertation. Robotics Institute Carnegie Mellon University.

[22] Siarry, P., Berthiau, G., Durbin, F. and Haussy, J. (1997) Enhanced simulated annealing for globally minimizing functions of many-continuous variables. *ACM Transactions on Mathematical Software*, **23** (2), 209–228.

[23] Blackburn, D., Miles, C., Wing, B. and Shepard, K. Biometrics Testing and Statistics, http://www.biometrics.gov/documents/biotestingandstats.pdf (last accessed Spetember 2014).

[24] Mansfield, A.J. and Wayman, J.L. (2002) Best Practices in Testing and Reporting Performance of Biometric Devices. NPL Report CMSC, 14/02, NPL publications.

[25] Baker, S. and Matthews, I. (2004) Lucas-Kanade 20 years on: a unifying framework. *International Journal of Computer Vision*, **56** (3), 221–255.

[26] ISO/IEC 19794-5 (2011) Information Technology - Biometric Data Interchange Formats - Part 5: Face Image Data, International Organization for Standardization.

[27] NIST. Face Vendor Recognition Test (FRVT) Homepage, http://www.nist.gov/itl/iad/ig/frvt-home.cfm (last accessed September 2014).

[28] Martinez, A.M. and Benavente, R. (1998) The AR Face Database. CVC Technical Report, 24. http://www.cvc.uab.es/

[29] UCSD. Yale Database, http://vision.ucsd.edu/content/yale-face-database (last accessed September 2014).

[30] Georghiades, A.S., Belhumeur, P.N. and Kriegman, D.J. (2001) From few to many: illumination cone models for face recognition under variable lighting and pose. *IEEE Transactions on Pattern Analysis and Machine Intelligence*, **23** (6), 643–660.

[31] CMU. CMU Multi-PIE database, http://www.multipie.org/ (last accessed: September 2014).

[32] SCface. Surveillance Cameras Face Database, http://www.scface.org/ (last accessed September 2014).

[33] Chen, T., Nwana, A., Kuan-Chuan and Chu, P.H. In Action (FIA) Face Video Database, http://amp.ece.cmu.edu/projects/FIADataCollection/ (last accessed September 2014).

[34] McGillFaces. McGill Real-World Face Video Database (McGillFaces), https://sites.google.com/site/meltemdemirkus/mcgill-unconstrained-face-video-database/ (last accessed September 2014).

[35] Vieira, T.F., Bottino, A., Laurentini, A. and De Simone, M. Siblings Database, http://areeweb.polito.it/ricerca/cgvg/siblingsDB.html (last accessed September 2014).

[36] Wolf, L., Hassner, T. and Maoz, I. YouTube Faces Database, http://www.cs.tau.ac.il/ wolf/ytfaces/ (last accessed September 2014).

[37] Zafeiriou, S., Atkinson, G.A., Hansen, M.F. *et al.* (2013) Face recognition and verification using photometric stereo: the photoface database and a comprehensive evaluation. *IEEE Transactions on Information Forensics and Security*, **8** (1), 121–135.

[38] 3DMAD. The 3D Mask Attack Database (3DMAD), https://www.idiap.ch/dataset/3dmad (last accessed September 2014).

[39] EURECOM KFD. The EURECOM Kinect Face Dataset (EURECOM KFD), http://rgb-d.eurecom.fr/ (last accessed September 2014).

[40] Grgic, M. and Delac, K. Face Recognition Datasets, http://www.face-rec.org/databases/ (last accessed September 2014).

5

Remote Sensing Image Registration in the Frequency Domain

5.1 Introduction

Many computer vision applications including tracking, video coding, geo-registration and 3D shape recovery are based on image registration. In this chapter, we will focus on satellite image registration, which is an important and challenging task due to the variety of data acquisition devices and sensors, size of the image data (hundreds of megapixels), the lack of proper validation algorithms and occlusions or interactions due to clouds and other atmospheric factors. Therefore, fast and efficient methodologies are required to provide reliable estimates of the transformations among the images with high accuracy. Satellite image registration is utilized in many fields for both research and industrial applications such as pattern recognition, change detection, fusion of multimodal data, climatology, surveillance, cartography, hydrology and geographical information systems.

The registration algorithms for satellite images can be separated into two main categories: the feature-based and the global or area-based techniques. Feature-based methods for image registration have been extensively studied in the literature [1]. In this category, the process of registration is usually performed in three main steps. Initially, we have the feature selection part extracting a possibly sparse set of distinct features. In the second step, each feature in a given image is compared with the corresponding features in a second image captured under different conditions. Pairs of features with similar characteristics are matched and are used as control points by establishing point-to-point correspondences between the two images. In the last step, based on the already estimated control points, a deformation model is obtained that corresponds to the transformation parameters between the given images. For more

Image, Video & 3D Data Registration: Medical, Satellite & Video Processing Applications with Quality Metrics,
First Edition. Vasileios Argyriou, Jesus Martinez Del Rincon, Barbara Villarini and Alexis Roche.
© 2015 John Wiley & Sons, Ltd. Published 2015 by John Wiley & Sons, Ltd.

details, an error measure that is based on the distances between the location of matched points is minimized so that a parametric geometric transformation that is assumed to relate the two images is recovered. In general, feature-based methods can be used for the recovery of very large camera motions.

Depending on the complexity of the feature selection process and their amount, the overall performance in terms of time and computational power varies. In general, the advantage of these algorithms is that it is not required to register the whole image allowing them to operate on higher resolutions, taking into consideration all the available details. Furthermore, in remote sensing applications, the feature-based registration approaches are separated to manual and automatic. Manual approaches could be applied in many registration problems, but this is not feasible when there is a large amount of data. In manual cases, the traditional approach to obtain control points is by human operators. Also, humans may have to classify and recognize image points on a map, and therefore, in this case, image registration becomes a tedious, difficult and costly process. Thus, automated techniques that require little or no human supervision are essential. In order to automate this process and increase the accuracy of global positioning systems (GPS), surveying is often involved, and more than one registration schemes are integrated into a system. Additionally, the user can provide information regarding the present features in the scene, the expected registration accuracy, the noise characteristics and other meta-data assisting further in this process. In the literature [2], it was shown that the best performance is provided by the scale-invariant feature transform (SIFT) in comparison with other descriptors, such as steerable filters [3], shape context [4], spin image [5] and differential invariants [6], which could also be used for spectral image registration [7, 8]. The principal component analysis SIFT (PCA-SIFT) and the speeded up robust features (SURF) techniques introduced in [9] and [10], respectively, try to provide similar accuracy with SIFT but with significantly less complexity and required processing time. Under the same principles, the work on nearest neighbour distance ratio (NNDR) [2] overcomes significant issues related to the complexity improving the overall speed. However, the above-mentioned feature-based methods may not be suitable for satellite image registration due to their large memory requirements [11], but are more suitable for 3D shape recovery techniques, which will be discussed in the following chapter. Regarding the correspondence problem between points or areas in two images, many methodologies were developed throughout the years. These methods include clustering [12], RANSAC [13–15], spectral matching [16–18], graph matching [19, 20] and relaxation labelling [21, 22]. These methods work well when the number of the control points is small, but their complexity increases significantly or break down when the provided point sets are large or have nonlinear geometric differences [23].

In contrast to feature-based methods, global methods for image registration estimate a geometric transformation between two images by using measurable image information, such as brightness variations or image cross-correlation measures, which is integrated from all the pixels [24]. Thus, global methods make use of all image information. Most global approaches can be classified as gradient-based and correlation-based [1]. Gradient methods accurately estimate motions in an iterative manner using non-linear optimization. Nevertheless, successful registration can be achieved only when a rough estimate of the unknown motion parameters is a priori available. Such estimates are usually provided by correlation schemes, which, in general, are better able to handle large motions. Thus, correlation methods usually provide an alternative to feature-based methods for the recovery of large motions when reliable feature matching is not feasible. In addition, very recently, a number of extensions have been proposed, which enable the very accurate estimation of motions to subpixel accuracy. Overall, correlation methods can be used for the estimation of both large and small motions. Finally, although correlation is a global method, it can be made local by dividing the input images into a number of smaller size subimages and processing each of them separately. According to this approach, in global methods [25, 26], a window of points or pixels in the first image is compared with the same-sized window in the second image. In this case, either the centres of the matched windows represent the control points or all the transformation parameters are directly obtained. About the estimation of accurate control points, methods based on invariant properties of image features [27–31], multiresolution matching [32–35] and the use of additional information such as digital terrain models (DTMs) [26] have been proposed in the literature over the last decades. Regarding the multiresolution methods, all areas in an image are treated in the same way but there are also cases such as the work in [23] that local image details could result to dynamic resolution changes.

Correlation methods are essentially useful for estimating translational shifts. Intuitively, to estimate the translational displacement between satellite images, a brute-force approach can be used: one examines all possible displacements and selects the one that yields the best match. Correlation is a method that runs an exhaustive search and uses the mean-squared error (MSE) criterion to perform matching. Direct computation of correlation is computationally expensive. Fortunately, we can compute correlation rapidly in the frequency (Fourier) domain using the fast Fourier transform (FFT). Therefore, correlation combines the merits of brute-force matching with the speed offered by FFT routines. For this very last reason, correlation-based motion estimation is usually coined as motion estimation in the Fourier domain. These methods utilize the Fourier transform properties for translation and rotational to estimate the transformation parameters [36, 37]. In this category, we have also wavelet-based global image registration approaches such as the approaches presented in [38–40]. Other examples include the work presented by Zhou *et al.* [38, 41] on serial and parallel strategies of wavelet-based automatic image registration: the method proposed by Chen *et al.* [42] on discrete multiwavelet transform (DMWT) and the approach based on the steerable Simoncelli filters that was proposed by Li *et al.* and Plaza in [43, 44]. The most common registration method in the frequency domain is phase correlation (PC) that was shown by Erives and Fitzgerald [45], and this class of methods will be the focus of this chapter.

5.2   Challenges in Remote Sensing Imaging

In remote sensing, image registration is the process of matching two different images or one image to a map in pixel or subpixel level. This registration process is required due to the multimodal source data, or due to the temporal variations and the viewpoint errors. Regarding the main challenges in satellite image analysis and registration, the different types of acquisition devices and their properties is an example. A list of instruments used for image acquisition is shown in Table 5.1, and it shows the variety in spatial and spectral resolution and in the supported subpixel mechanisms. The size of the captured data is also another significant challenge in registration, since the complexity and the required processing time may increase dramatically affecting the related applications that require near or real-time analysis. Additionally, in remote sensing, there is no known image model as it is in other areas such as face and human pose analysis. Therefore, the registration process cannot make any assumptions about the captured image data. Another challenge is related to the internal error of the acquisition device and the satellite navigation error that is linked to the obtained subpixel accuracy.

Furthermore, atmospheric and cloud interactions could affect the accuracy of the registration algorithm. An example of a satellite image with the smoke of a volcano in Japan occluding the surface is shown in Figure 5.1. Also, multitemporal and terrain/relief effects could influence significantly the registration process due to limitations in feature extraction and matching. Examples of these cases are shown in Figure 5.2, indicating the differences before and after natural hazards.

The accuracy of remote sensing image registration has major consequences in measuring global changes, since for example one pixel misregistration may result to 50% error in normalized difference vegetation index (NDVI) estimation using 250 m MODIS data. NDVI is an index used to identify vegetated areas and to detect green plant canopies in multispectral remote sensing data. Registration errors also may have resource management and economic and legal impact; therefore, the expected accuracy and efficiency is considerably high.

Table 5.1   A list of instruments used for remote sensing data capturing and their properties

Instrument	Satellite	Resolution	Subpixel
ASTER	Terra	15–90 m	Fit to surface
GOES	GOES I-M	1–8 km	Bi-section search
MISR	Terra	275 m	Least squares
MODIS	Terra	250 m-1 km	Fixed grid
HRS	SPOT	2.5 m	Not described
ETM+	Landsat-7	15–60 m	Fit to surface

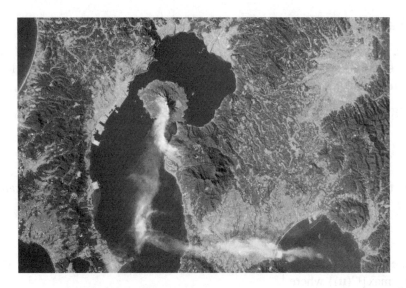

Figure 5.1 An example of atmospheric and cloud interactions. *Source*: Visible Earth NASA

Figure 5.2 Examples of multitemporal and terrain/relief effects due to natural hazards. *Source*: Earth Imaging Journal

5.3   Satellite Image Registration in the Fourier Domain

Let $I_i(\mathbf{x})$, $\mathbf{x} = [x, y]^T \in \mathcal{R}^2$, $i = 1, 2$ be two image functions. We denote by $\hat{I}_i(\mathbf{k})$, $\mathbf{k} = [k_x, k_y]^T \in \mathcal{R}^2$ the Cartesian FT of I_i and by M_i the magnitude of $\hat{I}_i$. Polar and log-polar Fourier representations refer to computing the FT as a function of $\mathbf{k}_p = [k_r, k_\theta]^T$ and $\mathbf{k}_l = [\log\ k_r, k_\theta]^T$, respectively, where $k_r = \sqrt{k_x^2 + k_y^2}$ and $k_\theta = \arctan(k_y/k_x)$.

5.3.1   Translation Estimation Using Correlation

Assume that we are given two images, I_1 and I_2, related by an unknown translation $\mathbf{t} = [t_x, t_y]^T \in \mathcal{R}^2$

$$I_2(\mathbf{x}) = I_1(\mathbf{x} + \mathbf{t}) \tag{5.1}$$

We can estimate $\mathbf{t}$ from the $2D$ cross-correlation function $C(\mathbf{u})$, $\mathbf{u} = [u, v]^T \in \mathcal{R}^2$ as $\hat{\mathbf{t}} = \arg_{\mathbf{u}} \max\{C(\mathbf{u})\}$ where [1]

$$C(\mathbf{u}) \triangleq I_1(\mathbf{u}) \star I_2(-\mathbf{u}) = \int_{\mathcal{R}^2} I_1(\mathbf{x})I_2(\mathbf{x} + \mathbf{u})d\mathbf{x} \tag{5.2}$$

From the *convolution theorem* of the FT [46], C can be alternatively obtained by

$$C(\mathbf{u}) = F^{-1}\left\{\hat{I}_1(\mathbf{k})\hat{I}_2^*(\mathbf{k})\right\} \tag{5.3}$$

where F^{-1} is the inverse FT and $\star$ denotes the complex conjugate operator. The *shift property* of the FT [46] states that if I_1 and I_2 are related by (5.1), then, in the frequency domain, it holds

$$\hat{I}_2(\mathbf{k}) = \hat{I}_1(\mathbf{k})e^{j\mathbf{k}^T\mathbf{t}} \tag{5.4}$$

and therefore (5.3) becomes

$$C(\mathbf{u}) = F^{-1}\left\{M_1^2(\mathbf{k})e^{-j\mathbf{k}^T\mathbf{t}}\right\} \tag{5.5}$$

The above analysis summarizes the main principles of frequency domain correlation-based translation estimation. For finite discrete images of size $N \times N$, correlation is efficiently implemented through (5.3), by zero padding the images to size $(2N - 1) \times (2N - 1)$ and using FFT routines to compute the forward and inverse FTs. If no zero padding is used, the match is cyclic and, in this case, the algorithm's complexity is $O(N^2 \log N)$.

---

[1] To be more precise, we assume hereafter that the images are of finite energy such that correlation integrals such as the one in (5.2) converge.

5.4 Correlation Methods

As it was shown in the previous section, image registration in the frequency domain is essentially a translation estimation problem, which can be rapidly solved using correlation. However, the standard definition of correlation in (5.2) is barely used in image processing. This is basically due to two reasons.

First, from (5.5), we observe that the phase difference term $e^{-jk^T t}$, which contains the translational information, is weighted by the magnitude M_1. In practice, in cases where (5.1) holds approximately only and $M_1 \neq M_2$, we estimate the translational displacement through (5.3). In this case, the phase difference function is weighted by the term $M_1 M_2$. Due to the low-pass nature of images, the weighting operation results in a correlation function with a number of broad peaks of large magnitude and a dominant peak whose maximum is not always located at the correct displacement.

Second, standard correlation is essentially a simplified version of the MSE criterion [46]. It is also well known that the MSE is not robust to outliers. Outliers are not unusual in real-world computer vision applications where, for example, data contain artefacts due to occlusions, illumination changes, shadows, reflections or the appearance of new parts/objects. For the above two reasons, a number of extensions to standard correlation have been proposed. In the following, we provide an overview of the most widely used methods with our focus mainly on the state-of-the-art gradient-based extensions to correlation.

Normalized correlation (NC) is a variant of standard correlation mostly known for its robustness to affine changes in image illumination. It is defined as

$$NC(\mathbf{u}) = \frac{\int_{R^2} \bar{I}_1(\mathbf{x}) \bar{I}_2(\mathbf{x} + \mathbf{u}) d\mathbf{x}}{\left(\int_{R^2} \bar{I}_1^2(\mathbf{x}) d\mathbf{x} \int_{R^2} \bar{I}_2^2(\mathbf{x} + \mathbf{u}) d\mathbf{x} \right)^{1/2}}, \tag{5.6}$$

where $\bar{I}_i(\mathbf{x}) = I_i(\mathbf{x}) - \int_{R^2} I_i(\mathbf{x}) d\mathbf{x}$. A fast implementation of NC is described in [47] while the FFT-based form of NC has been described only very recently in [48]. In the same work, a method to mask out pre-defined regions is also provided.

PC is perhaps the most widely used correlation-based method in image analysis. It looks for the maximum of the phase difference function, which is defined as the inverse FT of the normalized cross-power spectrum [49]

$$PC(\mathbf{u}) \triangleq F^{-1} \left\{ \frac{\hat{I}_2(\mathbf{k}) \hat{I}_1^*(\mathbf{k})}{|\hat{I}_2(\mathbf{k})| \|\hat{I}_1^*(\mathbf{k})|} \right\} = F^{-1} \{ e^{jk^T t} \} = \delta(\mathbf{u} - \mathbf{t}) \tag{5.7}$$

Thus, the resulting correlation function will be a 2D Dirac located at the unknown translation. This result is illustrated in Figure 5.3. The idea to use (5.7) comes naturally from the *shift property* of the FT and the importance of image phase, which typically captures most of the image salient features [50]. The method has two remarkable properties: excellent peak localization accuracy and immunity to uniform variations of illumination. In the presence of noise and dissimilar parts in the two

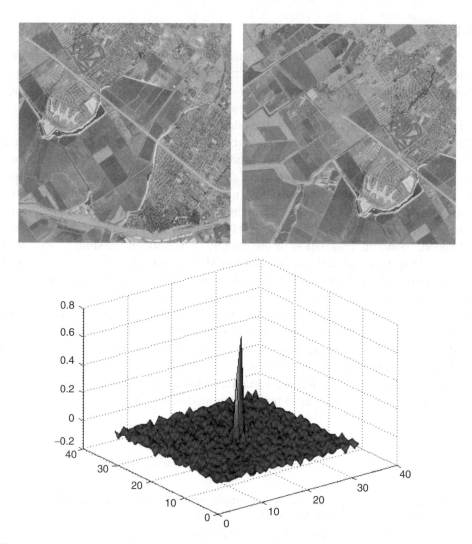

Figure 5.3 The correlation surface with a peak at the coordinates corresponding to the shift between the two pictures at the top. *Source*: Visible Earth database

images, however, the value of the peak is significantly reduced and the method may become unstable [51].

Gradient-based extensions to the baseline cross-correlation method have been proposed only recently and are considered state of the art. In [52], matching is performed by correlating the magnitudes of edge maps. Orientation correlation (OC) considers the orientation difference function solely [53]. Gradient correlation (GC) combines both magnitude and orientation information [54]. Robustness as key feature of gradient schemes is studied in [53]. It is attributed to Andrews' robust estimation kernel [55]. Based on the same principle, the authors in [56] describe a generic framework for fast robust correlation. In the following, we study the properties of gradient-based correlation schemes and present an alternative perspective that explains their robustness to outliers.

GC is defined as [54]

$$\text{GC}(\mathbf{u}) \triangleq G_1(\mathbf{u}) \star G_2^*(-\mathbf{u}) = \int_{\mathcal{R}^2} G_1(\mathbf{x}) G_2^*(\mathbf{x} + \mathbf{u}) d\mathbf{x} \tag{5.8}$$

where $G_i(\mathbf{x}) = G_{i,x}(\mathbf{x}) + jG_{i,y}(\mathbf{x})$ and $G_{i,x} = \nabla_x I_i$ and $G_{i,y} = \nabla_y I_i$ are the gradients along the horizontal and vertical directions, respectively.

Orientation correlation (OC) considers orientation information solely [53] by imposing

$$G_i(\mathbf{x}) \leftarrow \begin{cases} G_i(\mathbf{x})/|G_i(\mathbf{x})|, \text{ if } |G_i(\mathbf{x})| > \epsilon \\ 0, \text{ otherwise} \end{cases} \tag{5.9}$$

and ϵ is the value of a threshold. Thresholding $|G_i(\mathbf{x})|$ removes the contribution of pixels where gradient magnitude takes negligible values.

The robustness of GC to outliers can be shown using the polar representation of complex numbers. Let us define $R_i = \sqrt{G_{i,x}^2 + G_{i,y}^2}$ and $\Phi_i = \arctan G_{i,y}/G_{i,x}$. Based on this representation, GC takes the form [57]

$$\text{GC}(\mathbf{u}) = \int_{\mathcal{R}^2} R_1(\mathbf{x}) R_2(\mathbf{x} + \mathbf{u}) \cos[\Phi_1(\mathbf{x}) - \Phi_2(\mathbf{x} + \mathbf{u})] d\mathbf{x} \tag{5.10}$$

Equation (5.10) shows that GC is a joint metric that consists of two terms each of which can be an error metric itself. The first term is the correlation of the gradient magnitudes R_i. The magnitudes R_i reward pixel locations with strong edge responses and suppress the contribution of areas of constant intensity level, which do not provide any reference points for motion estimation. The second term is a cosine kernel applied on gradient orientations. This term is responsible for the Dirac-like shape of GC and its ability to reject outliers induced by the presence of dissimilar parts in the two images.

To show the latter point, let us first define the orientation difference function

$$\Delta\Phi_{\mathbf{u}}(\mathbf{x}) = \Phi_1(\mathbf{x}) - \Phi_2(\mathbf{x} + \mathbf{u}) \tag{5.11}$$

For a fixed $\mathbf{u} \neq \mathbf{t}$, we recall that the images do not match and, therefore, it is not unreasonable to assume that for any spatial location $\mathbf{x}_0 \in \mathcal{R}^2$, the difference in gradient orientation $\Delta\Phi_{\mathbf{u}}(\mathbf{x}_0)$ can take any value in the range $[0, 2\pi)$ with equal probability. Thus, for $\mathbf{u} \neq \mathbf{t}$, we assume that $\Delta\Phi_{\mathbf{u}}(\mathbf{x})$ is a stationary random process $y(t)$, with "time" index $t \triangleq \mathbf{x} \in \mathcal{R}^2$, which $\forall t$ follows a uniform distribution $U(0, 2\pi)$. If we define the random process $z(t) = \cos y(t)$, then it is not difficult to show that (under some rather mild assumptions) $\forall t$ the random variable $Z = z(t)$ satisfies

$$E(Z) \propto \int z(t) dt \equiv \int_{\mathcal{R}^2} \cos \Delta\Phi_{\mathbf{u}}(\mathbf{x}) d\mathbf{x} = 0, \quad \mathbf{u} \neq \mathbf{t} \tag{5.12}$$

which is essentially OC or alternatively GC after imposing $R_i = 1$, $i = 1, 2$.

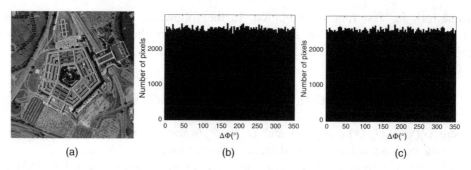

Figure 5.4 (a) The 512×512 pentagon image. (b) and (c) The distribution of the difference in orientation $\Delta\Phi$ between the original image and two circularly shifted versions. *Source:* http://commons.wikimedia.org/wiki/File:Pentagon-USGS-highres-cc.jpg

As an example, Figure 5.4 (a) shows the "Pentagon" image. We circularly shift the image in two different manners and, for each shift, we compute the difference $\Delta\Phi$ between the original and the shifted image. For each case, Figure 5.4 (b) and (c) shows the histogram with the distribution of $\Delta\Phi$. In both cases, $\Delta\Phi$ is well described by a uniform distribution and, therefore, the value of $\sum_i \cos \Delta\Phi(\mathbf{x}_i)$ will be approximately equal to zero.

Under the above assumptions, OC will be a Dirac function even when the given images match only partially. To show this, we model dissimilar parts by relaxing (5.1) as follows

$$I_1(\mathbf{x} + \mathbf{t}) = I_2(\mathbf{x}), \ \mathbf{x} \in \Omega \subseteq \mathcal{R}^2 \tag{5.13}$$

That is, after shifting I_1 by $\mathbf{t}$, I_1 and I_2 match only in $\mathbf{x} \in \Omega$. From the above analysis, we may observe that

$$OC(\mathbf{u})|_{\mathbf{u} \neq \mathbf{t}} = 0 \tag{5.14}$$

At $\mathbf{u} = \mathbf{t}$, we have

$$OC(\mathbf{t}) = \int_\Omega \cos \Delta\Phi_\mathbf{t}(\mathbf{x})d\mathbf{x} + \int_{\mathcal{R}^2 - \Omega} \cos \Delta\Phi_\mathbf{t}(\mathbf{x})d\mathbf{x} = \int_\Omega d\mathbf{x} \tag{5.15}$$

since $\Delta\Phi_\mathbf{t}(\mathbf{x}) = 0 \ \forall \ \mathbf{x} \in \Omega$ and in $\mathbf{x} \in \mathcal{R}^2 - \Omega$ the two images do not match. Overall, OC will be non-zero only for $\mathbf{u} = \mathbf{t}$, and its value at that point will be the contribution from the areas in the two images that match solely.

In the above analysis, it was assumed $R_i = 1$, $i = 1, 2$. To optimize the orientation difference function $\Delta\Phi$ of the image salient structures solely, the normalized gradient correlation (NGC) can be used [57]

$$NGC(\mathbf{u}) \triangleq \frac{G_1(\mathbf{u}) \star G_2^*(-\mathbf{u})}{|G_1(\mathbf{u})| \star |G_2(-\mathbf{u})|} = \frac{\int_{\mathcal{R}^2} G_1(\mathbf{x})G_2^*(\mathbf{x} + \mathbf{u})d\mathbf{x}}{\int_{\mathcal{R}^2} |G_1(\mathbf{x})G_2(\mathbf{x} + \mathbf{u})|d\mathbf{x}} \tag{5.16}$$

NGC has two interesting properties:

1. $0 \leqslant |NGC(\mathbf{u})| \leqslant 1$.
2. Invariance to affine changes in illumination.

The first property provides a measure to assess the correctness of the match. To show the second property, consider $I_2'(\mathbf{x}) = aI_2(\mathbf{x}) + b$ with $a \in \mathcal{R}^+$ and $b \in \mathcal{R}$. Then, by differentiation $G_2' = aG_2$; therefore, the brightness change due to b is removed. Additionally, $R_2' = aR_2$ and $\Delta\Phi_2' = \Delta\Phi_2$; thus the effect of the contrast change due to a will cancel out in (5.16). Note that if $a \in \mathcal{R}$, full invariance can be achieved by looking for the maximum of the absolute correlation surface.

5.5   Subpixel Shift Estimation in the Fourier Domain

Perhaps one of the most important applications of FFT-based correlation schemes is the estimation of subpixel shifts. This task is a typical pre-requisite for higher level applications in remote sensing registration related to image super-resolution and change detection.

Subpixel translation estimation in the Fourier domain is predominantly performed using PC [49]. This is basically due to two reasons. The first is directly related to the properties of PC itself. Subpixel extensions to PC [58–63] feature several remarkable properties: immunity to uniform variations of illumination, insensitivity to changes in spectral energy and excellent peak localization accuracy. Secondly, when *no aliasing occurs*, the estimation does not depend on the image content. This in turn largely simplifies the estimation process. Note however that when aliasing occurs, gradient-based correlation schemes may be preferred. The principles of FFT-based subpixel shift estimation are as follows.

Let $I(\mathbf{x})$, $\mathbf{x} = [x, y]^T \in \mathcal{R}^2$ be a continuous periodic image function with period $T_x = T_y = 1$. The Fourier series coefficients of I are given by

$$F_I(\mathbf{k}) = \int_\Omega I(\mathbf{x})e^{-j\omega_0\mathbf{k}^T\mathbf{x}}d\mathbf{x} \tag{5.17}$$

where $\Omega = \{\mathbf{x} : -1/2 \leq \mathbf{x} \leq 1/2\}$, $\mathbf{k} = [k, l]^T \in \mathcal{Z}^2$ and $\omega_0 = 2\pi$. If we sample I at a rate N with a 2D Dirac comb function $D(\mathbf{x}) = \sum_s \delta(\mathbf{x} - \mathbf{s}/N)$, we obtain a set of $N \times N$ discrete image values $I_1(\mathbf{m}) = I(\mathbf{m}/N)$, $\mathbf{m} = [m, n]^T \in \mathcal{Z}^2$ and $-N/2 \leq \mathbf{m} < N/2$. Using D, we can write the DFT of I_1 as

$$\hat{I}_1(\mathbf{k}) \triangleq \sum_\mathbf{m} I_1(\mathbf{m})e^{-j(2\pi/N)\mathbf{k}^T\mathbf{m}}$$

$$= \int_\Omega D(\mathbf{x})I(\mathbf{x})e^{-j2\pi\mathbf{k}^T\mathbf{x}}d\mathbf{x}$$

$$= \sum_\mathbf{s} \int_\Omega \delta(\mathbf{x} - \mathbf{s}/N)I(\mathbf{x})e^{-j2\pi\mathbf{k}^T\mathbf{x}}d\mathbf{x}$$

$$= \sum_{\mathbf{s}} F_I(\mathbf{k}) \star e^{-j2\pi \mathbf{k}^T \mathbf{s}/N}$$

$$= F_I(\mathbf{k}) \star \sum_{\mathbf{s}} e^{-j2\pi \mathbf{k}^T \mathbf{s}/N}$$

$$= F_I(\mathbf{k}) \star N^2 \sum_{\mathbf{s}} \delta(\mathbf{k} - \mathbf{s}N)$$

$$= N^2 \sum_{\mathbf{s}} F_I(\mathbf{k} - \mathbf{s}N) \tag{5.18}$$

where $-N/2 \leq \mathbf{k} < N/2$ and $\star$ denotes convolution.

Suppose we observe a shifted version $I(\mathbf{x} - \mathbf{x}_0)$ where $\mathbf{x}_0 = [x_0, y_0]^T, \{\mathbf{x}_0 : -1 < N\mathbf{x}_0 < 1\}$. From the Fourier series *shift* property, the corresponding coefficients are $F_I(\mathbf{k})e^{-j(2\pi/N)\mathbf{k}^T(N\mathbf{x}_0)}$. In a similar manner, we sample using D and get I_2. The DFT of I_2 is given by

$$\hat{I}_2(\mathbf{k}) = N^2 \sum_{\mathbf{s}} F_I(\mathbf{k} - \mathbf{s}N)e^{-j(2\pi/N)(\mathbf{k}-\mathbf{s}N)^T(N\mathbf{x}_0)} \tag{5.19}$$

If $F_I(\mathbf{k})$ is non-zero only for $-N/2 \leq \mathbf{k} < N/2$, then *no aliasing* occurs and we can write

$$\hat{I}_2(\mathbf{k}) = \hat{I}_1(\mathbf{k})e^{-j(2\pi/N)\mathbf{k}^T(N\mathbf{x}_0)} \tag{5.20}$$

Note that the well-known shift property of the DFT refers to integer shifts ($N\mathbf{x}_0 \in Z^2$) and does not assume aliasing-free signals. Hereafter, we assume that our sampling device eliminates aliasing.

Subpixel estimation with PC is based on the computation of the cross power spectrum

$$\hat{P}(\mathbf{k}) = e^{(2\pi/N)\mathbf{k}^T \Delta \mathbf{x}_0} \tag{5.21}$$

where $\Delta \mathbf{x}_0 = N\mathbf{x}_0$ can be estimated either in the spatial domain or in the frequency domain. In the former case, the inverse of (5.21) is first computed. As it was shown in [59], the inverse turns out to be the Dirichlet kernel (as opposed to the case of integer displacements that results in a Dirac function)

$$\text{PC}(\mathbf{m}) = \frac{\sin(\pi(\Delta \mathbf{x}_0 + \mathbf{m}))}{\sin(\pi(\Delta \mathbf{x}_0 + \mathbf{m})/N)} \tag{5.22}$$

which can be approximated by the sinc function. Finally, this approximation enables the analytic estimation of $\Delta \mathbf{x}_0$ [59].

Alternatively, frequency domain subpixel shift estimation uses linear regression to fit the phase values to a *2D* linear function (see Figure 5.5)

$$\mathbf{k}^T \mathbf{w} = \arg \hat{P}(\mathbf{k}) \tag{5.23}$$

Note that such an approach inevitably requires 2D phase unwrapping, which is a difficult ill-posed problem. To estimate the subpixel shifts, Stone *et al.* [58] propose to

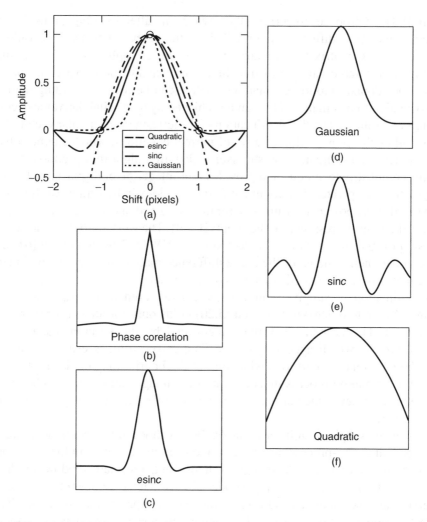

Figure 5.5 (a) Simple illustration of various fitted functions, (b) sample of the PC surface in the horizontal axis, (c) *esinc*-fitted function [65], (d) Gaussian-fitted function, (e) sinc-fitted function and (f) quadratic-fitted function

perform the fitting after masking out frequency components corrupted by aliasing. Masking is based on two criteria: (i) distance from highest peak of the PC surface and (ii) amplitude relative to a threshold. The latter is dynamically determined as follows. Frequencies are sorted by magnitude and are progressively eliminated starting with the lowest. The authors claim that there exists a range in which the accuracy of the computed motion estimates becomes stable and independent of the degree of progressive elimination. This range indirectly determines the required threshold. An extension to the method for the additional estimation of planar rotation has been proposed in [64].

In [60], Hoge proposes to perform the unwrapping after applying a rank-1 approximation to the phase difference matrix. This method is based on the observation that a 'noise-free' PC matrix (i.e. a matrix between shifted replicas of the same image) is a rank-1, separable-variable matrix. In a non-ideal situation, that is a 'noisy' PC matrix, the subpixel motion estimation problem can be recast as finding the rank-1 approximation to that matrix. This can be achieved by using singular value decomposition (SVD) followed by the identification of the left and right singular vectors. These vectors allow the construction of a set of normal equations, which can be solved to yield the required estimate. The subpixel registration algorithm proposed by Hoge is a low complexity technique and provides non-integer pixel displacements without interpolation. It operates directly in the frequency domain by fitting two lines on the normalized cross-power spectrum, one for the x-axis and one for the y-axis. The slope of each fitted line corresponds to the subpixel shift. The work in [62] is a noise-robust extension to [60], where noise is assumed to be AWGN. The authors in [61] derive the exact parametric model of the phase difference matrix and solve an optimization problem for fitting the analytic model to the noisy data.

The estimation of subpixel shifts in the Fourier domain is largely affected by aliasing. As we have shown, Equation (5.20) holds only for aliasing-free frequency components. The problem is further aggravated by the spectral leakage caused by edge effects, which appear when the Fourier spectrum is estimated using the FFT. Phase correlation performs whitening of the Fourier magnitude spectra, which makes the estimation process independent of the image content and mathematically well-defined; however, the normalization does not offer any advantages when dealing with aliasing.

To estimate subpixel shifts accurately, PC methods rely on efficient pre- and post-processing. Pre-processing typically includes the use of window functions to reduce the boundary effect. A strategy to choose an appropriate window for all cases does not exist. The choice is usually a trade-off between the number of image features preserved and the amount of aliasing removed. Post-processing strategies include subspace approximations [60] and masking out specific frequency components [58, 60]. While the subspace decomposition is straightforward to apply, the masking operation poses difficulties. Reasonable approaches include preserving frequency components, which lie within a radius $r = c_1(N/2)$ of the spectrum origin, and discarding components with magnitude less than a threshold c_2. The values of c_1 and c_2 are chosen arbitrarily, which makes the estimation process rather unstable.

The above problems can be alleviated by kernel-based extensions to GC [66]. These methods extend GC to achieve subpixel accuracy. They are based on modelling the dominant singular vectors of the 2D GC matrix with a generic kernel, which is derived by studying the structure of GC assuming natural image statistics. As it was shown in [66], these extensions are better able to cope with edge and aliasing effects for two reasons. First, they are based on grey-level edge maps and, therefore, discontinuities due to periodization will appear only if very strong edges exist close to the image boundaries. In practice, such edge effects can be efficiently suppressed by using a window with a relatively small cut-off parameter (e.g. a Tukey with cut-off 0.25), which removes a very small amount of the signal energy. Second, they replace masking with a weighting operation. In particular, the phase difference function $e^{-j(2\pi/N)\mathbf{k}^T(N\mathbf{x}_0)}$, which contains the translational information, is weighted by a filtered power spectrum. This filter is the product of two terms. The first term is a band-pass filter. This filter gradually filters out low-frequency components, which do not provide any reference points for registration, as well as reduces the effect of high-frequency noise and aliasing. The second term puts additional emphasis on those frequency components that correspond to image salient features such as edges and corners.

5.6   FFT-Based Scale-Invariant Image Registration

In this section, we will focus on global image registration schemes that make use of all image information for the estimation of translations, rotations and isotropic scalings in images. This set of image transformations is also known as the class of similarity transforms.

For the class of similarity transforms, a frequency domain approach to image registration capitalizes on the *shift property* of the Fourier transform (FT) and the use of FFT routines for the rapid computation of correlations. More specifically, the FT *shift property* states that translations (shifts) in the image domain correspond to linear phase changes in the Fourier domain leaving the Fourier magnitudes unaffected. Rotations and scalings in images result in rotations and scalings of the corresponding FT magnitudes by the same and inverse amounts, respectively [67]. Thus, we can capitalize on the *shift* property to reduce the problem of searching a 4D parameter space (two dimensions for translation, one for rotation and one for scaling) to one of searching two 2D subspaces (one for translation and one for rotation and scaling). To estimate the unknown parameters using correlation, we must additionally reformulate the search problem as a translation estimation one. This is possible by using the log-polar representation of the FT magnitudes, which reduces rotation and scaling to a 2D translation [51].

Assume that we are given two images, I_1 and I_2, related by a translation $\mathbf{t}$, rotation $\theta_0 \in [0, 2\pi)$ and scaling $s > 0$, that is

$$I_2(\mathbf{x}) = I_1(D\mathbf{x} + \mathbf{t}) \qquad (5.24)$$

where $D = s\Theta$ and $\Theta = \begin{bmatrix} \cos\theta_0 & \sin\theta_0 \\ -\sin\theta_0 & \cos\theta_0 \end{bmatrix}$. In the Fourier domain, it holds [67]

$$\hat{I}_2(\mathbf{k}) = (1/|\Delta|)\hat{I}_1(\mathbf{k}')e^{j\mathbf{k}'^T\mathbf{t}} \qquad (5.25)$$

where

$$\mathbf{k}' = D^{-T}\mathbf{k} \qquad (5.26)$$

and Δ is the determinant of D. Taking the magnitude in both parts of (5.25) and substituting $D^{-T} = \Theta/s$, $\Delta = s^2$ gives

$$M_2(\mathbf{k}) = (1/|\Delta|)M_1(\mathbf{k}') = (1/s^2)M_1(\Theta\mathbf{k}/s) \tag{5.27}$$

Using the log-polar representation gives (ignoring $1/s^2$) [51]

$$M_2(\mathbf{k}_l) = M_1(\mathbf{k}_l - [\log s, \theta_0]^T) \tag{5.28}$$

We can observe that in the log-polar Fourier magnitude domain, the rotation and scaling reduce to a 2D translation, which can be estimated using correlation. After compensating for the rotation and scaling, we can recover the remaining translation using correlation in the spatial domain. Note that if $\tilde{\theta}_0$ is the estimated rotation, then it is easy to show that $\tilde{\theta}_0 = \theta_0$ or $\tilde{\theta}_0 = \theta_0 + \pi$. To resolve the ambiguity, one needs to compensate for both possible rotations, compute the correlation functions and, finally, choose as valid solution to the one that yields the highest peak [51].

Although the estimation of rotation and scaling appears to be straightforward, in practice, it turns out to be a rather difficult task. An obvious reason for that is that rotation and scaling induce large non-overlapping regions between images while (5.24) holds only for the overlapping regions. Because the FT is a global transform applied to the whole image, it immediately follows that (5.27) holds only approximately. Unfortunately, this is not the only reason that hinders the estimation of rotation and scaling in the frequency domain. Other important but less obvious problems are discussed in detail as follows.

First, the estimation process is hindered by low-frequency components induced by areas of constant intensity level. To illustrate this, consider the images in Figure 5.6(a) and (c) and the scenario where the motion is purely rotational (see Figure 5.6(b) and (d)). To estimate the rotation, suppose that we use the 1D representation $A(k_\theta) = \int M(k_r, k_\theta)dk_r$ and correlation over the angular parameter k_θ. As we may observe, the image contains a wide range of frequencies and, consequently, A is almost flat (Figure 5.6(e), dashed line). In this case, matching by correlation can be unstable.

Second, conversion from Cartesian to polar/log-polar induces much larger interpolation error for low-frequency components. Figure 5.7 clearly illustrates the problem. We may observe that near the origin of the Cartesian, gridless data are available for interpolation. It is also evident that for Cartesian-to-log-polar conversion, the situation becomes far more problematic since the log-polar representation is extremely dense near the origin. Thus, recently proposed DFT schemes [68–70] sample the FT on non-Cartesian grids, which geometrically are much closer to the polar/log-polar ones. To recover the rotation and scaling, the method in [68] relies on the pseudopolar FFT [71], which rapidly computes a discrete FT on a nearly polar grid. The pseudopolar grid serves as an intermediate step for a log-polar Fourier representation, which is obtained using nearest-neighbour interpolation. Overall, the total accumulated interpolation error is decreased; nevertheless, the pseudopolar FFT is not a true polar Fourier representation and the method estimates the rotation and scaling in an iterative manner. In [69], the authors propose to approximate the log-polar

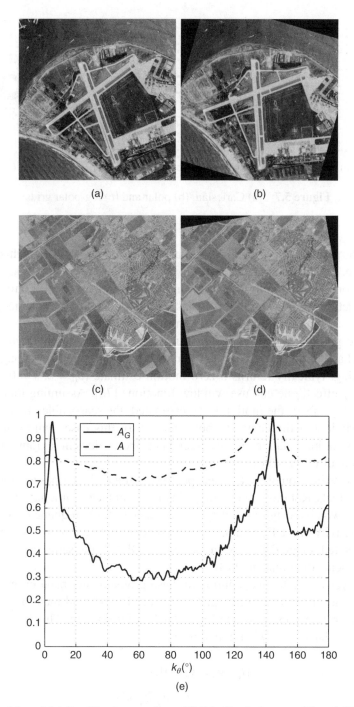

Figure 5.6 (a) and (c) Satellite images from Visible Earth dataset; (b) and (d) the rotated versions; and (e) the 1D representations A (dashed line) and A_G (solid line). *Source*: Visible Earth database

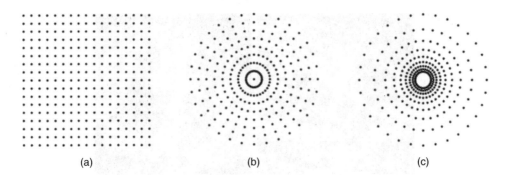

Figure 5.7   (a) Cartesian, (b) polar and (c) log-polar grids

DFT by interpolating the pseudo-log-polar FFT. Compared to [68], this method is non-iterative. Finally, the main idea in [70] is to obtain more accurate log-polar DFT approximations by efficiently oversampling the lower part of the Fourier spectrum using the fractional FFT. Overall, in all these methods, accuracy is enhanced, however, at the cost of additional computational complexity.

Third, the periodic nature of the FFT induces boundary effects, which result in spectral leakage in the frequency domain. Attempting to register images with no pre-processing, typically returns a zero-motion estimate ($\theta_0 = 0$, $s = 0$). To reduce the boundary effect, one can use window functions [72]. Assuming that there is no prior knowledge about the motion to be estimated, the reasonable choice is to place the same window at the centre of both images. In this case, windowing not only results in loss of information but also attenuates pixel values in regions shared by the two images in different ways.

Fourth, the estimation of the Fourier spectrum using FFT routines is largely affected by aliasing effects. Rotations and scalings in images induce additional sources of aliasing artefacts, which are aggravated by the presence of high frequencies. For example, the commutativity of the FT and image rotation does not hold in the discrete case: the DFT of a rotated image differs from the rotated DFT of the same image resulting in rotationally depending aliasing [73].

A simple yet very effective solution to these problems has been proposed in [57]. To estimate the rotation and scaling, I_i is replaced by G_i and then M_{G_i} is used as a basis to perform correlation in the log-polar Fourier domain. This is possible since from (5.26) we have $k_r = sk'_r$ and therefore

$$
\begin{aligned}
M_{G_2}(\mathbf{k}) &= k_r M_2(\mathbf{k}) \\
&= (1/s)k'_r M_1(\mathbf{k}') \\
&= (1/s)M_{G_1}(\mathbf{k}') \\
&= (1/s)M_{G_1}(\Theta\mathbf{k}/s)
\end{aligned}
\tag{5.29}
$$

In contrast to M, M_{G_i} captures the frequency response of the image salient features solely. With respect to the example of Figure 5.6, A_G, which is obtained by averaging M_G, efficiently captures possible directionality of the image salient features: the two main orientations of the edges in the image give rise to two distinctive peaks in A_G (Figure 5.6(e), solid line). Using A_G to perform correlation, matching will be more accurate and robust. Second, by using M_{G_i}, there is no need to resort to sophisticated non-Cartesian FFT methods. In fact, as it was shown in [57], standard Cartesian FFT and bilinear interpolation without oversampling is sufficient. This is because the undesirable effect of low-frequency components is largely eliminated if the representation M_{G_i} is used. This comes naturally since the bottom line from the example in Figure 5.6 is that discarding low frequencies from the representation will also result in more robust and accurate registration. Also, there is no need to use image windowing. This is because discontinuities due to periodization will appear only if very strong edges exist close to the image boundaries. Finally, using filters with band-pass spectral selection properties to compute G_i effectively reduces the effect of high-frequency noise and aliasing in the estimation process.

5.7 Motion Estimation in the Frequency Domain for Remote Sensing Image Sequences

Motion estimation in remote sensing image sequences is the process by which the temporal relationship between two successive frames is determined. When an object in a 3D environment moves, the luminance of its projection in 2D is changing either due to non-uniform lighting or due to motion. Assuming uniform lighting, the changes can only be interpreted as object movement. Understanding how objects move helps us track changes in an efficient way. Thus, the aim of motion estimation techniques is to accurately model the motion field.

In a typical image sequence, there is no 3D information about the contents of the objects. The 2D projection approximating a 3D scene is known as a 'homography', and the velocity of the 3D objects corresponds to the velocity of the luminance intensity on the 2D projection, known as 'optical flow'. Another term is the 'motion field', a 2D matrix of motion vectors, corresponding to how each pixel or block of pixels moves. General 'motion field' is a set of motion vectors, and this term is related to the 'optical flow' term, with the latter being used to describe dense 'motion fields'.

Motion estimation for change detection in satellite images has been studied over the last years. In this section, motion detection algorithms based on the Fourier transform will be analysed including methods operating on both fixed size and adaptive blocks, on object shapes or on pixel level. In block matching techniques, both the reference and the current frame are split into non-overlapping blocks. These algorithms assume that within each block, motion is equal. One motion vector is assigned to all pixels within a block; thus, all the pixels perform the same motion and the algorithm searches the neighbouring area of the reference frame. PC method is an example of motion estimation algorithm that can be applied on each block.

A lot of work has been done in the area of adaptive block size motion estimation, and many quad-tree algorithms have been proposed. A fast and highly accurate approach of this category is quad-tree PC described earlier. Note that PC could be also replaced by gradient-based correlation schemes.

5.7.1   Quad-Tree Phase Correlation

Baseline PC operates on a pair of images or, more commonly (i.e. for applications requiring a measure of local motion), a pair of co-sited rectangular blocks f_t and f_{t+1} of identical dimensions belonging to consecutive frames or fields of a moving sequence sampled, respectively, at t and $t + 1$. The estimation of motion relies on the detection of the maximum of the cross-correlation function between f_t and f_{t+1}, [74].

The starting point for a quad-tree partition of frame f_t involves a global PC operation between f_t and f_{t+1}. We assume that f_{t+1} is the target frame (the frame whose motion we seek to estimate) and f_t is a reference frame (typically, the current frame in an ordered sequence). This global operation yields a measure of global motion by identification of the maximum (highest peak) of the resulting correlation surface $PC_{t,t+1}$ as expressed by (5.7).

5.7.1.1   Partition Criteria

The next step is to examine whether target frame f_{t+1} should be partitioned or not to four quadrants $\{f^i_{t+1}\}$ $i = 1, \dots, 4$. This is determined by applying suitable criteria. The first criterion is derived from the identification of dominant peaks on the PC surface and is computationally less demanding. The second criterion is based on the computation of the motion-compensated prediction error, requires more computations but guarantees a monotonic decrease in the error with an increasing number of iterations.

5.7.1.2   Peak Ratio

At this point, a key property of $PC_{t,t+1}$ needs to be revisited; namely that, in addition to the highest peak, which corresponds to the primary motion component between f_t and f_{t+1}, secondary peaks also correspond to other dominant motion components. Another key property, which is very relevant to our purposes, is that the height of each of these peaks is a good measure of reliability of the corresponding motion component. For example, high and sharp peaks can be relied upon to correspond to real

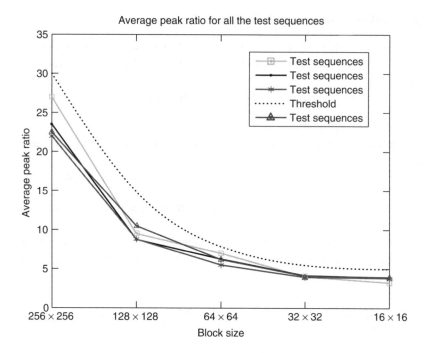

Figure 5.8   Average peak ratio for all the test sequences

motion, while low and/or blunt peaks may be the result of noise or cross talk. Experimental evidence suggests that it is at most the highest two to three peaks that can be relied upon. These properties can be taken into account to formulate a sensible quad-tree partition criterion. Assuming a ranking of peaks according to height, that is $\left\{ PC_{t,t+1}^{m}, PC_{t,t+1}^{m-1}, PC_{t,t+1}^{m-2}, \ldots \right\}$ where $PC_{t,t+1}^{m}$ is the highest peak, $PC_{t,t+1}^{m-1}$ the second highest and so on, we use the ratio of the highest to the second highest peak, that is

$$r = \frac{PC_{t,t+1}^{m}}{PC_{t,t+1}^{m-1}} \tag{5.30}$$

as a quad-tree partition criterion. If the ratio is higher relative to a threshold, we assume that there is only one dominant motion component in the scene and the algorithm does not proceed any further, otherwise the target frame is partitioned into four quadrants. It has been found that the ratio threshold depends on block size. As the block size becomes smaller, the ratio of the two highest peaks reduces exponentially. In Figure 5.8, the average peak ratio for different block sizes ranging from 256×256 to 16×16 pixels is shown. Therefore, the threshold depends on block size, and the best results were obtained using settings higher by 15–25% relative to the average ratio (for a given block size).

5.7.1.3   Motion-Compensated Prediction Error

This criterion uses the translational motion parameters $(k_m + dx, l_m + dy)$ obtained by (5.20), (5.21) and (5.22) to form a motion-compensated prediction $\hat{f}_{t+1}$ of f_{t+1} using f_t, that is

$$\hat{f}_{t+1}(x, y) = f_t(x + k_m + dx, y + l_m + dy) \qquad (5.31)$$

where (x, y) are pixel locations.

A motion-compensated prediction is also formed for each of the four quadrants as follows:

$$\hat{f}^i_{t+1}(x, y) = f^i_t(x + k^i_m + dx^i, y + l^i_m + dy^i) i = 1, \ldots , 4 \qquad (5.32)$$

where $k^i_m + dx^i, l^i_m + dy^i)$ are subpixel accurate motion parameters for each of the four quadrants.

Finally, if the mean-squared motion-compensated prediction error (MSE) of the split is lower than the MSE before the split, which is equivalent to the following holding true:

$$\sum_{i=1}^{4} \sum_{x,y} [\hat{f}^i_{t+1}(x, y) - f^i_{t+1}(x, y)]^2 < \sum_{x,y} [\hat{f}_{t+1}(x, y) - f_{t+1}(x, y)]^2 \qquad (5.33)$$

then the target image/block is allowed to split into four quadrants. This criterion guarantees a monotonic decrease in the motion-compensated prediction error with an increasing number of iterations. While this is an attractive feature, it should be noted that it incurs a higher computational cost relative to the first criterion.

5.7.1.4   Correlating Unequal Size Image Blocks

Assuming that the target frame has been partitioned as earlier, motion parameters need to be estimated for each of the resulting four quadrants. One obvious course of action would have been to partition the reference frame f_t in a similar manner, that is to four quadrants $\{f^i_{t+1}\}$ $i = 1, \ldots , 4$ and perform PC between co-sited quadrants f^i_t and f^i_{t+1}. However, this would have restricted the range of motion parameters accordingly, that is inside the ith quadrant. A consequence of this would be that the motion parameters of a fast-moving object traversing quadrant boundaries during a single frame period would be impossible to estimate. For this reason, it is preferable to correlate quadrant f^i_{t+1} in the target frame with the entire reference frame f_t. Nevertheless, this has the obvious disadvantage of requiring a PC operation to be performed between two images of unequal sizes, that is the reference being four times larger than the target. One straightforward way to go round this problem would be to increase the size of the target image by a factor of 2 in each dimension by zero-order (constant-value) extrapolation, that is by symmetric (top-bottom, left-right) insertion of zeros or mid-grey values to the unknown pixel locations outside f^i_{t+1} until the latter assumes equal dimensions to f_t. Furthermore, this can be implemented in the frequency domain by interpolative upsampling of F_{t+1}^i, that is the Fourier transform of f_{t+1}^i. We have used bilinear interpolation to obtain $\tilde{F}_{t+1}^i$, whose dimensions are now identical to F_t, hence allowing the PC operation to be implementable without any further modifications. Interpolative upsampling in the frequency domain has the obvious practical advantage that the Fourier transform of the target image requires far less computations than otherwise.

5.7.1.5 Further Iterations

The algorithm proceeds in an iterative manner to determine whether or not each quadrant f_{t+1}^i will be partitioned any further. We will assume that each subsequent partition is denoted by an additional index, that is $\{f_{t+1}^{i,j,k,\cdots}\}$, $i,j,k,\ldots = 1,\ldots,4$ so that the cardinality of the superscript set is equal to the levels of partitions carried out. For example, $\{f_{t+1}^{i,j}\}$ denotes the jth quadrant (at the second level of partition) of the ith quadrant (at the first level of partition). Due to the fact that some quadrants may occasionally fail the criteria and hence remain undivided, not all $\{f_{t+1}^{i,j,k,\cdots}\}$ will exist for all possible combinations of $i,j,k,\ldots = 1,\ldots,4$. Using the above notation, during a second iteration of the algorithm, quadrant f_{t+1}^i is further partitioned to four subquadrants $\{f_{t+1}^{i,j}\}, j = 1,\ldots,4$. Reference frame f_t is also partitioned to four quadrants $\{f_t^i\}$, $i = 1,\ldots,4$ and target subquadrant $f_{t+1}^{i,j}$ is phase-correlated with its co-sited parent quadrant f_t^i according to above-mentioned section.

The algorithm can be allowed to proceed to a natural termination point, that is until all subquadrants contain no more than a single motion component or until blocks can no longer provide the minimum area of support required for PC to work properly. Experimental evidence suggests that this point is typically reached for 16×16 pixel blocks.

5.7.2 Shape Adaptive Motion Estimation in the Frequency Domain

In [75], a shape adaptive PC algorithm was presented using padding techniques. Hill and Vlachos suggested the use of conventional extrapolation, that is by padding with either zeros (zero-padding) or the average (mean) intensity of the arbitrary-shaped object. The use of the mean instead of zeros for padding reduces the artificial edges caused by the segmentation and padding, resulting in better and more accurate estimates. For two successive frames f_1 and f_2 of a sequence, the steps of the motion estimation algorithm proposed by Hill are listed as follows and an example is illustrated in Figure 5.9.

- Shape segmentation and extraction.
- Zero- or Mean-padding.
- PC.

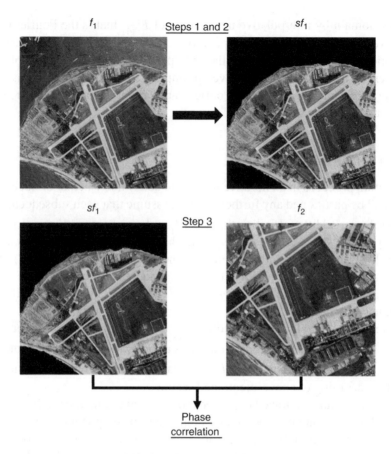

Figure 5.9   Steps of shape adaptive PC algorithm using padding techniques. *Source*: Visible Earth database

This method provides accurate results especially when more than one moving objects are present in the scene. On the other hand, one of the main drawbacks is the fact that even the use of the mean value for padding cannot eliminate the artificial edges created by extrapolation. Additionally, the possibility to have a good match at the padded area (background) is not negligible, resulting in wrong and inaccurate estimates. In order to overcome these issues, GC can be applied since no artificial edges are created during the extrapolation.

5.7.3   Optical Flow in the Fourier Domain

A primary problem in image sequence analysis is the measurement of optical flow or image velocity. As mentioned earlier, the objective is the estimation of a 2D motion field (the projection of 3D velocities) from image intensity. The problem with optical flow estimation originates from the fact that many assumptions of the world are violated in practice. Two main constrains are usually used: data conservation and spatial consistency. The former expresses the fact that the intensity structure of most regions is preserved in time. This assumption is obviously violated on motion boundaries and when shadows or specular reflections are present. Additionally, it does not stand for occlusions, transparency effects, noise and other distortions. Spatial coherence assumes that neighbouring pixels are likely to belong to the same object and therefore should result in similar displacements. The presence of multiple simultaneously moving objects and the complexity of the scene motion that cannot be described by the model being used (e.g. rotation is modelled as translation) tend to violate the spatial coherence assumption. The consequence of these violations is significant errors in the optical flow estimation [76, 77].

In this section, a brief review and description of well-known dense motion estimation algorithms operating in the frequency domain is presented.

5.7.3.1 Phase-Based Methods

Thomas in [78] suggests a vector measurement method that has the accuracy of the Fourier techniques described earlier, combined with the ability to measure multiple objects' motion in a scene. The method is also able to assign vectors to individual pixels if required.

In the first stage, the input frames are divided into fairly large blocks (64×64 pixels or even bigger) and a PC is performed between corresponding blocks in successive frames, resulting in a number of correlation surfaces describing the movement present in different areas of the frame. Each correlation surface is searched to locate several dominant peaks resulting from the motion of the objects within each block. As a result, several motion vectors could be measured by each correlation process, providing a list of motion vectors likely to be present in the scene.

The second stage of the process involves the assignment of the possible vectors measured in the first stage to appropriate areas of the frame. This can be achieved by applying motion compensation and calculating the motion-compensated prediction error for each estimated motion vector. This process produces error surfaces, equal in number to vectors obtained at the first stage, indicating how well the motion vector fits each area of the frame. The vector resulting in the lowest error is assigned to each area. Areas where sufficiently low error cannot be produced by motion vectors correspond to erratic motion or uncovered background. Therefore, a different motion estimation algorithm should be used for such areas or additional motion vectors from neighbouring areas could be attempted.

The conceptual basis of the method is the trial of a limited number of motion vectors obtained in the PC stage. It eliminates the necessity to assign vectors on a block-by-block basis and minimizes the area of the frame used to determine if a vector fits or not, since the number of trial vectors is limited.

5.7.3.2   Wavelet-Based Registration

Wavelet pyramidal decomposition was recently introduced and suggested for satellite image registration due to its inherent multiresolution properties [79]. According to these methods, the image is recursively decomposed into four sets of coefficients (LL, HL, LH, HH) by filtering the image with a low-pass and a high-pass filter. Also, methods for image registration that combine both feature-based and global-based matching operating in the wavelet domain were proposed [80]. Features extracted in the wavelet domain combined with normalized cross-correlation matching are used to estimate the control points reducing the problems due to noise and other local distortions.

5.8   Evaluation Process and Related Datasets

Considering the impact of remote sensing image registration on sociopolitical and industrial decisions, the performance evaluation mechanisms are essential. The ranking process of the registration algorithms could be based on different criteria and the needs of the related application. Therefore, the performance requirements are set during the design of the system in terms of accuracy, robustness and efficiency. For example, in some cases, the accuracy may be more important than the evaluation time or vice versa in a real-time application.

Traditionally, the assessment of satellite image registration algorithms is based on 'global similarity' of the resultant images (compensated images) and the 'inverse consistency' of the transformation [81, 82]. The metrics related to the first approach assume that the motion-compensated (transformed) image is similar to the reference image. High similarity is assumed to indicate good registration, although not fully guaranteed [83]. Regarding the inverse registration consistency, it is quantified with an error between a forward and reverse transformation. 'Forward transformation' is described as compensating the target image to match the reference, and 'reverse transformation' is the opposite. This metric is based on the fact that the 'forward'-compensated image should be identical to the 'inverse of the reverse' image.

Examples of global similarity metrics are the L_2-norm that minimizes the sum of squared differences (SSD) of overlapped windows

$$\text{SSD}(x, y) = \sum_{m=0}^{M-1} \sum_{n=0}^{N-1} [I_1(m, n) - I_2(m - x, n - y)]^2 \qquad (5.34)$$

Also, a metric that maximizes the normalized cross-correlation (NCC) is utilized:

$$\text{NCC}(x, y) = \frac{\sum_{m=0}^{M-1} \sum_{n=0}^{N-1} [I_1(m, n) - \bar{I}_1][I_2(m - x, n - y) - \bar{I}_2]}{\sqrt{\sum_{m=0}^{M-1} \sum_{n=0}^{N-1} [I_1(m, n) - \bar{I}_1]^2 \sum_{m=0}^{M-1} \sum_{n=0}^{N-1} [I_2(m - x, n - y) - \bar{I}_2]^2}}$$

$$(5.35)$$

Another global similarity metric that maximizes the statistical dependence between the given images is mutual information (MI)

$$MI(I_1, I_2) = \sum_a \sum_b p_{I_1,I_2}(a, b) \log \left(\frac{p_{I_1,I_2}(a, b)}{p_{I_1}(a)p_{I_2}(b)} \right) \qquad (5.36)$$

where a and b are pixel intensities and p_I indicates the probability of a given intensity value being present in the image I.

PC can be considered also as a similarity measure that is based on the location of the peak in the correlation surface (closer to $(0, 0)$ indicates higher accuracy), and on the height of the peak or the ratio of the highest peak versus the second highest peak.

Partial Hausdorff distance (PHD) is another similarity metric given by the following equation

$$PHD(I_1, I_2) = K^{th}_{p_1 \in I_1} \min_{p_2 \in I_2} dist(p_1, p_2) \qquad (5.37)$$

and based on the Hausdorff distance, two images are close if every pixel of either image is close to some pixels of the other image.

Finally, discrete Gaussian mismatch (DGM) is the similarity measure and is defined as

$$DGM(I_1, I_2) = 1 - \frac{\sum_{a \in I_1} w_\sigma(a)}{|I_1|} \qquad (5.38)$$

where $w_\sigma(a) = \exp\left(-\frac{dist(a,I_2)^2}{2\sigma^2}\right)$ denotes the weight of the point a.

Regarding, the assessment of the registration algorithms based on the inverse consistency is given by the root-mean-squared error (RMSE) between the forward and reverse motion-compensated images.

5.8.1   Remote Sensing Image Datasets

The amount of datasets for remote sensing image analysis is limited with the most well known to include the Visible Earth (NASA) database, the IKONOS multispectral set, the Vision Image Archive Large image database from University of Massachusetts, the Modis Airborne simulator dataset and the USC-SIPI image database.

In some cases, synthetic data could be used for the evaluation of the registration methods since the groundtruth is available. These images can be synthesized based on the Lambertian model or other bidirectional reflectance distribution functions used for computer-generated imaginary and flight simulators (see Figure 5.10).

5.9   Conclusion

The research issues reported in this chapter are focused on different categories of remote sensing image registration, starting with the challenges of satellite image registration and moving to the most well-known translation estimation methods using

Figure 5.10   Examples of simulated images for remote sensing registration [84]

PC. The concept of subpixel shift estimation in the Fourier domain is introduced and efficient algorithms are presented. Approaches to provide scale and rotation-invariant image registration based on FFT are discussed, providing detailed methodologies with further applications on image sequences. Example cases for motion estimation in the Fourier domain based on quad-tree, shape adaptive and optical flow are provided. Finally, metrics for the registration evaluation and related datasets are presented, indicating the significance of remote sensing image registration assessment on sociopolitical and industrial issues.

5.10   Exercise – Practice

In this exercise, the estimation of translation, rotation and scaling parameters between two given images will be performed. Find the rotation, scaling and translation between the given images in figure 5.6, using the methods discussed in this chapter based on the Fourier properties. So, moving in the frequency domain, perform PC to estimate the translation; in order to obtain the rotation and scaling parameters, polar and log-polar representations are required based on the steps analysed in Section 5.6.

A final step in this exercise will be to improve the previous parts for translation, rotation and scaling estimations by including subpixel accuracy based on the concepts discussed in Section 5.5.

References

[1] Zitova, B. and Flusser, J. (2003) Image registration methods: a survey. *Image and Vision Computing*, **21**, 977–1000.

[2] Mikolajczyk, C.A. and Schmid, K. (2005) A performance evaluation of local descriptors. *IEEE Transactions on Pattern Analysis and Machine Intelligence*, **27**, 1615–1630.

[3] Freeman, W. and Adelson, E. (1991) The design and use of steerable filters. *IEEE Transactions on Pattern Analysis and Machine Intelligence*, **13**, 891–906.

[4] Belongie, S., Malik, J. and Puzicha, J. (2002) Shape matching and object recognition using shape contexts. *IEEE Transactions on Pattern Analysis and Machine Intelligence*, **24**, 509–522.

[5] Lazebnik, S., Schmid, C. and Ponce, J. (2003) A sparse texture representation using affine-invariant regions. *Proceedings of the IEEE Computer Society Conference on Computer Vision and Pattern Recognition*, Vol. 2, pp. 319–324.

[6] Koenderink, J. and van Doorn, A. (1987) Representation of local geometry in the visual system. *Biological Cybernetics*, **55**, 367–375.

[7] Lowe, D. (1999) Object recognition from local scale-invariant features. *Proceedings of the IEEE Computer Society Conference on Computer Vision and Pattern Recognition*, Vol. **2**, pp. 1150–1157.

[8] Lowe, D. (2004) Distinctive image features from scale-invariant keypoints. *International Journal of Computer Vision*, **60**, 91–110.

[9] Ke, Y. and Sukthankar, R. (2004) Pca-sift: a more distinctive representation for local image descriptors. *Proceedings of the IEEE Computer Society Conference on Computer Vision and Pattern Recognition*, Vol. **2**, pp. 506–513.

[10] Bay, H., Ess, A., Tuytelaars, T. and Van Gool, L. (2008) Speeded-up robust features (SURF). *Computer Vision and Image Understanding*, **9** (110), 346–359.

[11] Lee, I., Seo, D.-C. and Choi, T.-S. (2012) Entropy-based block processing for satellite image registration. *Entropy*, **14** (12), 2397–2407.

[12] Stockman, G., Kopstein, S. and Benett, S. (1982) Matching images to models for registration and object detection via clustering. *IEEE Transactions on Pattern Analysis and Machine Intelligence*, **4** (3), 229–241.

[13] Chum, O. and Matas, J. (2008) Optimal randomized RANSAC. *IEEE Transactions on Pattern Analysis and Machine Intelligence*, **30** (8), 1472–1482.

[14] Fischler, M.A. and Bolles, R.C. (1981) Random sample consensus: a paradigm for model fitting with applications to image analysis and automated cartography. *Communications of the ACM*, **24** (6), 381–395.

[15] Matas, J. and Chum, O. (2004) Randomized RANSAC with T(d,d) test. *Image and Vision Computing*, **22** (10), 837–842.

[16] Carcassoni, M. and Hancock, E.R. (2003) Spectral correspondence for point pattern matching. *Pattern Recognition*, **36**, 193–204.

[17] Scott, G.L. and Longuet-Higgins, H.C. (1991) An algorithm for associating the features of two images. *Proceedings of the Royal Society of London Series B*, **244**, 21–26.

[18] Shapiro, L.S. and Brady, J.M. (1992) Feature-based correspondence: an eigenvector approach. *Image and Vision Computing*, **10** (5), 283–288.

[19] Bolles, R.C. (1979) Robust feature matching through maximal cliques. „*SPIE Conference Imaging Applications for Automated Industrial Inspection and Assembly*, Vol. **182**, pp. 140–149.

[20] Goshtasby, A. and Stockman, G.C. (1985) Point pattern matching using convex hull edges. *IEEE Transactions on Systems, Man, and Cybernetics*, **15** (5), 631–637.

[21] Cheng, F.-H. (1996) Point pattern matching algorithm invariant to geometrical transformation and distortion. *Pattern Recognition Letters*, **17**, 1429–1435.

[22] Zhao, J., Zhou, S., Sun, J. and Li, Z. (2010) Point pattern matching using relative shape context and relaxation labeling. *Proceedings of the 2nd International Conference on Advanced Computer Control (ICACC)*, Vol. **5**, pp. 516–520.

[23] Wu, Z. and Goshtasby, A. (2012) Adaptive image registration via hierarchical Voronoi subdivision. *IEEE Transactions on Image Processing*, **21** (5), 2464–2473.

[24] Irani, M. and Anandan, P. (2000) *All About Direct Methods, Lecture Notes in Computer Science*, Vol. **1883**, Springer-Verlag, pp. 267–277.

[25] Vila, M., Bardera, A., Feixas, M. and Sbert, M. (2011) Tsallis mutual information for document classification. *Entropy*, **13**, 1694–1707.

[26] Rignot, E.J.M., Kowk, R., Curlander, J.C. and Pang, S.S. (1991) Automated multisensor registration: requirements and techniques. *Photogrammetric Engineering and Remote Sensing*, **57** (8), 1029–1038.

[27] Flusser, J. and Suk, T. (1994) A moment-based approach to registration of images with affine geometric distortion. *IEEE Transactions on Geoscience and Remote Sensing*, **32**, 382–387.

[28] Li, H., Manjunath, B. and Mitra, S. (1995) A contour-based approach to multisensor image registration. *IEEE Transactions on Image Processing*, **4**, 320–334.

[29] Wang, W. and Chen, Y. (1997) Image registration by control points pairing using the invariant properties of line segments. *Pattern Recognition Letters*, **18**, 269–281.

[30] Dai, X. and Khorram, S. (1999) A feature-based image registration algorithm using improved chain-code representation combined with invariant moments. *IEEE Transactions on Geoscience and Remote Sensing*, **37**, 2351–2362.

[31] Sharp, G., Lee, S. and Wehe, D. (2002) ICP registration using invariant features. *IEEE Transactions on Pattern Analysis and Machine Intelligence*, **24**, 90–102.

[32] Djamdji, J., Bijaoui, A. and Maniere, R. (1993) Geometrical registration of images: the multiresolution approach. *Photogrammetric Engineering and Remote Sensing*, **59** (5), 645–653.

[33] Zhang, Z., Zhang, J., Liao, M. and Zhang, L. (2000) Automatic registration of multi-source imagery based on global image matching. *Photogrammetric Engineering and Remote Sensing*, **66** (5), 625–629.

[34] McGuire, M. and Stone, H. (2000) Techniques for multiresolution image registration in the presence of occlusions. *IEEE Transactions on Geoscience and Remote Sensing*, **38**, 1476–1479.

[35] Jacp, J.-J. and Roux, C. (1995) Registration of 3d images by genetic optimization. *Pattern Recognition Letters*, **16**, 823–841.

[36] Xu, M. and Varshney, P.K. (2009) A subspace method for fourier-based image registration. *IEEE Geoscience and Remote Sensing Letters*, **6** (3), 491–494.

[37] Li, Q., Sato, I. and Sakuma, F. (2008) A novel strategy for precise geometric registration of GIS and satellite images. *IEEE International Geoscience and Remote Sensing Symposium*, Vol. **2**, No. 2, pp. 1092–1095.

[38] Zhou, H., Yang, X., Liu, H. and Tang, Y. (2005) First evaluation of parallel methods of automatic global image registration based on wavelets. *International Conference on Parallel Processing*, pp. 129–136.

[39] Hong, G. and Zhang, Y. (2005) The image registration technique for high resolution remote sensing image in hilly area. *International Society of Photogrammetry and Remote Sensing Symposium*.

[40] Hong, G. and Zhang, Y. (2007) Combination of feature-based and area-based image registration technique for high resolution remote sensing image. *IEEE International Geoscience and Remote Sensing Symposium*, pp. 377–380.

[41] Zhou, H., Tang, Y., Yang, X. and Liu, H. (2007) Research on grid-enabled parallel strategies of automatic wavelet based registration of remote-sensing images and its application chinagrid. *4th International Conference on Image and Graphics*, pp. 725–730.

[42] Chen, C.-F., Chen, M.-H. and Li, H.-T. (2007) *Fully Automatic and Robust Approach for Remote Sensing Image Registration, CIARP LNCS Springer*, Vol. **4756**, Springer-Verlag, pp. 891–900.

[43] Li, Q., Sato, I. and Murakami, Y. (2007) Steerable filter based multiscale registration method for JERS-1 SAR and ASTER images. *IEEE International Geoscience and Remote Sensing Symposium*, pp. 381–384.

[44] Plaza, A., Le Moigne, J. and Netanyahu, N.S. (2005) Automated image registration using morphological region of interest feature extraction. *International Workshop on the Analysis of Multi-Temporal Remote Sensing Images*, pp. 99–103.

[45] Erives, H. and Fitzgerald, G.J. (2006) Automatic subpixel registration for a tunable hyperspectral imaging system. *IEEE Geoscience and Remote Sensing Letters*, **3** (3), 397–400.

[46] Gonzalez, R. and Woods, R. (2002) *Digital Image Processing*, 2nd edn, Pearson Education Asia Pte Ltd, Singapore.

[47] Lewis, J.P. (1995) Fast normalized cross-correlation. *Vision Interface*, **95**, 120–123.

[48] Padfield, D. (2010) Masked FFT registration. *IEEE Conference on Computer Vision and Pattern Recognition, 2010. CVPR 2008*, IEEE, pp. 1–8.

[49] Kuglin, C. and Hines, D. (1975) The phase correlation image alignment method. Proceedings of the *IEEE Conference on Cybernetics and Society*, pp. 163–165.

[50] Oppenheim, A. and Lim, J. (1981) The importance of phase in signals. *Proceedings of the IEEE*, **69** (5), 529–541.

[51] Reddy, B. and Chatterji, B. (1996) An FFT-based technique for translation, rotation, and scale-invariant image registration. *IEEE Transactions on Image Processing*, **5** (8), 1266–1271.

[52] Glasbey, C. and Martin, N. (1996) Multimodality microscopy by digital image processing. *Journal of Microscopy*, **181**, 225–237.

[53] Fitch, A., Kadyrov, A., Christmas, W. and Kittler, J. (2002) Orientation correlation. *BMVC*, pp. 133–142.

[54] Argyriou, V. and Vlachos, T. (2003) Estimation of sub-pixel motion using gradient cross-correlation. *Electronic Letters*, **39** (13), 980–982.

[55] Huber, P.J. (1981) *Robust Statistics*, John Wiley & Sons, Inc., New York.

[56] Fitch, A., Kadyrov, A., Christmas, W. and Kittler, J. (2005) Fast robust correlation. *IEEE Transactions on Image Processing*, **14** (8), 1063–1073.

[57] Tzimiropoulos, G., Argyriou, V., Zafeiriou, S. and Stathaki, T. (2010) Robust FFT-based scale-invariant image registration with image gradients. *IEEE Transactions on Pattern Analysis and Machine Intelligence*, pp. 1899–1906.

[58] Stone, H.S., Orchard, M., Chang, E.-C. and Martucci, S. (2001) A fast direct fourier-based algorithm for subpixel registration of images. *IEEE Transactions on Geoscience and Remote Sensing*, **39** (10), 2235–2243.

[59] Foroosh, H., Zerubia, J. and Berthod, M. (2002) Extension of phase correlation to sub-pixel registration. *IEEE Transactions on Image Processing*, **11** (2), 188–200.

[60] Hoge, W. (2003) Subspace identification extension to the phase correlation method. *IEEE Transactions on Medical Imaging*, **22** (2), 277–280.

[61] Balci, M. and Foroosh, H. (2006) Subpixel estimation of shifts directly in the fourier domain. *IEEE Transactions on Image Processing*, **15** (7), 1965–1972.

[62] Keller, Y. and Averbuch, A. (2007) A projection-based extension to phase correlation image alignment. *Signal Processing*, **87**, 124–133.

[63] Ren, J., Vlachos, T. and Jiang, J. (2007) Subspace extension to phase correlation approach for fast image registration. *Proceedings of IEEE ICIP*, pp. 481–484.

[64] Vandewalle, P., Susstrunk, S. and Vetterli, M. (2006) A frequency domain approach to registration of aliased images with application to super-resolution. *EURASIP Journal on Applied Signal Processing*, **2006**, 1–14.

[65] Argyriou, V. and Vlachos, T. (2006) A study of sub-pixel motion estimation using phase correlation. *Proceedings of the BMVC*, Vol. **1**, pp. 387–396.

[66] Tzimiropoulos, G., Argyriou, V. and Stathaki, T. (2011) Subpixel registration with gradient correlation. *IEEE Transactions on Image Processing*, **20** (6), 1761–1767.

[67] Bracewell, R.N., Chang, K.-Y., Jha, A.K. and Wang, Y.-H. (1993) Affine theorem for two-dimensional fourier transform. *Electronics Letters*, **29** (3), 304.

[68] Keller, Y., Averbuch, A. and Israeli, M. (2005) Pseudopolar-based estimation of large translations, rotations and scalings in images. *IEEE Transactions on Image Processing*, **14** (1), 12–22.

[69] Liu, H., Guo, B. and Feng, Z. (2006) Pseudo-log-polar fourier transform for image registration. *IEEE Signal Processing Letters*, **13** (1), 17–21.

[70] Pan, W., Qin, K. and Chen, Y. (2009) An adaptable-multilayer fractional fourier transform approach for image registration. *IEEE Transactions on Pattern Analysis and Machine Intelligence*, **31** (3), 400–413.

[71] Averbuch, A., Donoho, D., Coifman, R. and Israeli, M. (2007) Fast slant stack: a notion of radon transform for data in cartesian grid which is rapidly computable, algebraically exact, geometrically faithful and invertible. *IEEE International Conference on Image Processing*, **1** (1), 57–60.

[72] Harris, F. (1978) On the use of windows for harmonic analysis with the discrete fourier transform. *Proceedings of the IEEE*, **66** (1), 51–83.

[73] Stone, H., Tao, B. and MacGuire, M. (2003) Analysis of image registration noise due to rotationally dependent aliasing. *Journal of Visual Communication and Image Representation*, **14**, 114–135.

[74] Argyriou, V. and Vlachos, T. (2007) Quad-tree motion estimation in the frequency domain using gradient correlation. *IEEE Transactions on Multimedia*, **9** (6), 1147–1154.

[75] Hill, L. and Vlachos, T. (2001) Motion measurement using shape adaptive phase correlation. *Electronics Letters*, **37** (25), 1512.

[76] Horn, B.K. (1986) *Robot Vision*, MIT Press McGraw-Hill.

[77] Verri, A. and Poggio, T. (1987) Against quantitative optical flow. *Proceedings of the 1st International Conference on Computer Vision*, pp. 171–180.

[78] Thomas, G. (1987) Television motion measurement for DATV and other applications. BBC Res. Dept. Rep., No. 1987/11, BBC Res.

[79] Moigne, J. (1994) Parallel registration of multi-sensor remotely sensed imagery using wavelet coefficients. *Proceedings of the SPIE: Wavelet Applications*, Vol. **2242**, pp. 432–443.

[80] Hong, G. and hang, Y. (2007) Combination of feature-based and area-based image registration technique for high resolution remote sensing image. *IEEE international symposium on Geoscience and remote sensing, IGARSS 2007*, Vol. **2242**, pp. 377–380.

[81] Gerçek, D., Çe?meci, D., Güllü, M.K. *et al.* (2013) Accuracy improvement and evaluation measures for registration of multisensor remote sensing imagery. *Journal of Geodesy and Geoinformation*, **1** (2), 97–103.

[82] Christensen, G.E., Geng, X., Kuhl, J.G. *et al.* (2006) Introduction to the non-rigid image registration evaluation project (NIREP). *Proceedings of the 3rd International Conference on Biomedical Image Registration*, Springer-Verlag, pp. 128–135.

[83] Zitová, B. and Flusser, J. (2003) Image registration methods: a survey. *Image and Vision Computing*, **21**, 977–1000.

[84] Carlotto, M.J. and Stein, M.C. (1990) A method for searching for artificial objects on planetary surfaces. *Stein Journal of the British Interplanetary Society*, **43**, 209–216.

6

Structure from Motion

6.1 Introduction

One of most significant and popular problems in computer vision is structure from motion (SfM) and is of great interest in many fields with tremendous applications. It is focused on generating 3D models and structures of real scenes and objects from 2D images that can be used in robotics for navigation, tracking, entertainment (e.g. video games and movies), augmented reality, mosaics, reverse engineering, security and architecture. Modelling of a scene or an object in 3D for these applications can be performed using laser scan systems or manually using design tools. Either these approaches are time consuming or the cost is high for the related applications. SfM techniques allow scenes and models to be automatically reconstructed from simple image acquisition devices, providing an efficient solution to the above problems.

SfM is defined as the process of estimating a 3D structure of rigid objects and the relative camera motion from 2D image sequences without knowing the external camera parameters. Regarding the rigidity constrain, it can be relaxed in some cases and also the internal camera parameters may not be available but still the estimation of the 3D structures is possible. In general, SfM is a dual problem with two separate parts, with one of them aiming to obtain the unknown structure of an object given the camera location over time and the second trying to estimate the camera location or motion from known points in the 3D space.

The state-of-the-art methods for the reconstruction of static 3D scenes can be separated into categories based on the related assumptions. In the first category, image features and their correspondences over time are considered known, therefore the algorithms try to estimate the camera movement and the 3D locations of the selected features. This approach is used in applications related to SfM [1–4] and to automatic calibration of cameras and projectors [5]. In the second category, we have methods for stereo [6] and volumetric reconstruction [7–8], assuming that the camera viewpoints are given. In the last category, the shape of a model is provided and the camera

Image, Video & 3D Data Registration: Medical, Satellite & Video Processing Applications with Quality Metrics,
First Edition. Vasileios Argyriou, Jesus Martinez Del Rincon, Barbara Villarini and Alexis Roche.
© 2015 John Wiley & Sons, Ltd. Published 2015 by John Wiley & Sons, Ltd.

viewpoints over each image are estimated [9–10]. Therefore, accurate and reliable camera calibration, and estimation of feature correspondences based on registration techniques are essential in SfM. The main issues with these approaches are the errors in feature-tracking and their propagation over the following frames. In order to reduce their effect, methods that reject outliers were suggested [11–13].

The most common pipeline used in SfM consists of a set of subsystems shown in Figure 6.1. In this chapter, we provide an overview of this process analysing and reviewing each step. Initially, the image sequence is acquired containing an object or scene. Since the observed scene is the same as the camera is moving, single features of that scene are expected to be present in more than one frame. These features, usually points of high curvature, are extracted and tracked through the captured frames. In order to calculate the SfM, multiple view geometry is applied on these features. Additionally, the external camera parameters are estimated based on the extracted features, and finally the 3D structure of these points is obtained.

In the following sections, the camera pinhole model is reviewed and camera calibration methods are analysed. The problem of identifying the corresponding points is discussed, and concepts such as projection matrix, stereo geometry and triangulation that are essential to recover the 3D structure of a scene or a given object are presented.

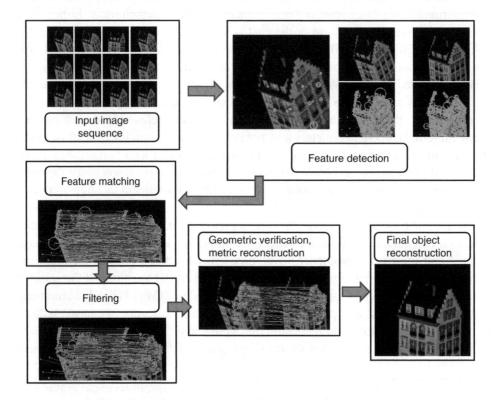

Figure 6.1   The main structure of the pipeline used in Structure from Motion

The state-of-the-art reconstruction methods based on SfM for more than two views are reviewed. Finally, the simultaneous localization and mapping (SLAM) method is discussed and well-known datasets and metrics, which are used for the evaluation of SfM techniques, are analysed.

6.2 Pinhole Model

During the image acquisition process, the captured images are 2D projections of the real-world scenes consisting of geometric (e.g. positions, points, lines) and photometric information (e.g. colour, intensity, specularity). Therefore, there are complex 3D to 2D relationships and many camera models available in the literature (e.g. pinhole camera, orthographic, scaled orthographic, perspective projection) approximating these relationships. The camera pinhole projection model is one of the most well-known models that offer reasonable approximation of most real cameras. This model provides a relationship between a point in 3D space and its corresponding coordinates on a 2D image. The concept of pinhole camera is based on placing a barrier in between the acquired object and the film to block off most of the rays, allowing only few to go through the opening known as the aperture reducing any blurring effects (see Figure 6.2).

The steps that are required to perform this transformation are analysed later. Initially, the extrinsic camera parameters are defined describing the camera orientation and position in the 3D world. If the camera's centre of projection is not at the centre of the world coordinates (0, 0, 0) and also the orientation is not perpendicular to the image plane but arbitrary, then a transformation to the origin of the coordinate system is essential. Therefore, a 4×4 transformation matrix M that relates the points in the world coordinate system $P = [x \quad y \quad z \quad 1]$ to locations in the camera coordinate system is defined as

$$\begin{bmatrix} X & Y & Z & 1 \end{bmatrix}^T = \begin{bmatrix} R & T \\ 0 & 1 \end{bmatrix} \begin{bmatrix} x & y & z & 1 \end{bmatrix}^T = M \begin{bmatrix} x & y & z & 1 \end{bmatrix}^T \tag{6.1}$$

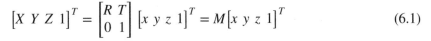

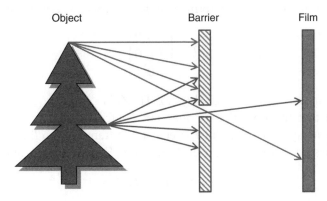

Figure 6.2   A representation of the concept of a pinhole camera

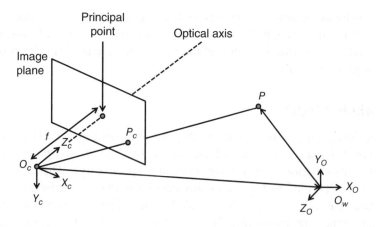

Figure 6.3 The camera orientation and position in the camera coordinate system

where R is a rotation 3×3 matrix and T is a 1×3 vector representing the camera orientation and position of the origin O_w of the world in the camera coordinate system (see Figure 6.3).

Regarding the intrinsic camera parameters, it is assumed that the focal length is equal to f, indicating the distance between the camera's centre of projection O_c and the image plane (principal point). Given a 3D point P and the transformation matrix M, it is imaged on the image plane at coordinate $P_c = [u \quad v]$. In order to find the location of P_c, an approach based on similar triangles is used,

$$f/Z = u/X = v/Y \tag{6.2}$$

which gives

$$u = fX/Z \tag{6.3}$$

$$v = fY/Z \tag{6.4}$$

This can be written in a matrix form using homogeneous coordinates for P_c as

$$(u \quad v \quad w)^T = \begin{pmatrix} f & 0 & 0 \\ 0 & f & 0 \\ 0 & 0 & 1 \end{pmatrix} (x \quad y \quad z) \tag{6.5}$$

Generating the point $P_c = (u, v, w) = (fX/Z, fY/Z, 1)$.

In the case that the optical axis does not coincide with the origin of the 2D image coordinate system, the point P_c needs to be translated to the desired location (i.e. origin). Assuming that (t_u, t_v) is the translation, Equations (6.3) and (6.4) become

$$u = fX/Z + t_u \tag{6.6}$$

$$v = fY/Z + t_v \tag{6.7}$$

and the matrix form is given by

$$(u \; v \; w)^T = \begin{pmatrix} f & 0 & t_u \\ 0 & f & t_v \\ 0 & 0 & 1 \end{pmatrix} (x \; y \; z) \tag{6.8}$$

If we want to take into consideration the camera resolution and express the location of the point P_c in pixels, Equations (6.6–6.8) need to be updated. Given that the pixels are rectangle and that their horizontal and vertical resolutions are r_x and r_y, respectively, we obtain the pixel coordinates of P_c as

$$u = r_u fX/Z + r_u t_u \tag{6.9}$$

$$v = r_v fY/Z + r_v t_v \tag{6.10}$$

and expressing them in a matrix form, we have

$$(u \; v \; w)^T = \begin{pmatrix} r_u f & s & r_u t_u \\ 0 & r_v f & r_v t_v \\ 0 & 0 & 1 \end{pmatrix} (x \; y \; z) \Rightarrow P_c' = DP_c \tag{6.11}$$

with C to depend on the focal length, the principal axis and all the intrinsic camera parameters. One more parameter that can be considered is the skew s, with Equation (6.11) becoming

$$C = \begin{pmatrix} r_u f & s & r_u t_u \\ 0 & r_v f & r_v t_v \\ 0 & 0 & 1 \end{pmatrix} \tag{6.12}$$

with matrix C called intrinsic parameter matrix (or camera calibration matrix).

Combining the extrinsic and intrinsic parameters from Equations (6.1) and (6.12), we obtain the projected point P_c from P as

$$P_c = C[R \; T]P = DP \tag{6.13}$$

where D is the 3×4 camera projection matrix. In the following section, the estimation of the calibration matrix will be discussed.

6.3   Camera Calibration

In order to extract information for the 3D world from 2D images, camera calibration is an essential part of this process. This task of camera calibration is focused on estimating the extrinsic and intrinsic parameters of the acquisition device that affect the imaging process. These parameters include the centre of the image plane (i.e. principal point), the focal length, pixel scaling factors, skew and other distortions.

Significant work has been done in this area [14–19], and different approaches have been considered.

One of the most well-known approaches is based on capturing images of a controlled scene using a calibration object. According to this techniques [20–24], the image locations of known 3D points are extracted (see Figure 6.4). Given a point $P_i = [x_i \ y_i \ z_i]^T$ in 3D space and its image position $P_{c_i} = [u_i \ v_i]^T$, two equations are obtained due to this correspondence from Equation (6.13).

$$u_i = \frac{p_{11}X_i + p_{12}Y_i + p_{13}Z_i + p_{14}}{p_{31}X_i + p_{32}Y_i + p_{33}Z_i + p_{34}} \tag{6.14}$$

$$v_i = \frac{p_{21}X_i + p_{22}Y_i + p_{23}Z_i + p_{24}}{p_{31}X_i + p_{32}Y_i + p_{33}Z_i + p_{34}} \tag{6.15}$$

Rearranging these two equations in a matrix form that must be satisfied by the 12 unknown elements of the projection matrix D, we obtain

$$\begin{bmatrix} X_1 & Y_1 & Z_1 & 1 & 0 & 0 & 0 & 0 & -u_1X_1 & -u_1Y_1 & -u_1Z_1 & -u_1 \\ 0 & 0 & 0 & 0 & X_1 & Y_1 & Z_1 & 1 & -v_1X_1 & -v_1Y_1 & -v_1Z_1 & -v_1 \\ \vdots & \vdots & \vdots & \vdots & \vdots & \vdots & \vdots & \vdots & & \vdots & & \vdots \\ X_n & Y_n & Z_n & 1 & 0 & 0 & 0 & 0 & -u_nX_n & -u_nY_n & -u_nZ_n & -u_n \\ 0 & 0 & 0 & 0 & X_n & Y_n & Z_n & 1 & -v_nX_n & -v_nY_n & -v_nZ_n & -v_n \end{bmatrix} \begin{bmatrix} p_{11} \\ p_{12} \\ p_{13} \\ p_{14} \\ p_{21} \\ p_{22} \\ p_{23} \\ p_{24} \\ p_{31} \\ p_{32} \\ p_{33} \\ p_{34} \end{bmatrix} = 0 \tag{6.16}$$

Figure 6.4   The camera orientation and position in the camera coordinate system

Without considering the scaling parameter, there are 11 unknowns; since n calibration points provide $2n$ equations, 6 points in the 3D space are required at least to estimate the calibration matrix of the camera.

In order to solve this system, a least squares solution can be utilized. Therefore, we rewrite the equations in a matrix form as

$$Lp = 0 \tag{6.17}$$

where p is a 12×1 vector with the unknown elements of the calibration matrix, L is the $2n \times 12$ matrix with the estimated correspondences of the n 3D points and the projected ones. The numerical solution is given by applying singular value decomposition (SVD) [25] on the L matrix.

$$L = U\Sigma V^T \tag{6.18}$$

where Σ consists of the singular values and the eigenvectors are stored in V. According to the work of Abdel-Aziz and Karara in [26], the constrain $p_{34} = 1$ was placed to avoid a trivial solution of all the unknowns being equal to zero. Later Faugeras and Toscani [27] suggested the constrain $p_{31}^2 + p_{32}^2 + p_{33}^2 = 1$ to avoid problems due to singularity when the correct value of p_{34} is close to zero.

The matrix p with the 12 parameters does not have a physical meaning, and therefore, the estimation of these values does not complete the calibration process. Therefore, techniques to retrieve some of the physical camera parameters have been proposed in the literature. In [28, 29] an approach based on RQ, decomposition was proposed by Melen allowing the estimation of 11 physical parameters using the following decomposition.

$$L = \lambda Q^{-1} B^{-1} FRT \tag{6.19}$$

where λ is the scaling factor and matrices Q, B and F provide the principal point (u_0, v_0), the distortion coefficients (a_1, a_2) and the focal length f, respectively.

$$Q = \begin{bmatrix} 1 & 0 & -u_0 \\ 0 & 1 & -v_0 \\ 0 & 0 & 1 \end{bmatrix}, \quad B = \begin{bmatrix} 1+a_1 & a_2 & 0 \\ a_2 & 1-a_1 & 0 \\ 0 & 0 & 1 \end{bmatrix}, \quad F = \begin{bmatrix} f & 0 & 0 \\ 0 & f & 0 \\ 0 & 0 & 1 \end{bmatrix} \tag{6.20}$$

The matrices R and T define the rotation and translation from the object coordinate system to the camera coordinate system. Matrix L is used to adjust error in orthogonality.

6.4 Correspondence Problem

The process of identifying features (points) in two or more images captured from different points of view that derive from the same feature (point) in the 3D space is known as the correspondence problem. The proposed solutions to this problem can be separated into two categories: the manual methods that require human interaction and the automatic methods. Regarding the first category, a person identifies the corresponding points or geometric primitives interactively in each captured image [30, 31]. The main issue related to these manual approaches is the required time to manually process all the given images and the accuracy of the user.

The automatic methods try to find the correspondence of all the pixels between images captured from different viewpoints, or a smaller number of points (features) are used. The approaches that estimate correspondences of all the pixels based on optical flow are considered in the cases that the displacement of the camera between the viewpoints is small, while the feature-based methods are used when the baseline is large. In general, the techniques based on feature correspondences provide more accurate estimates having many applications compared to the optical flow ones [32–34]. In these methods, interest points are extracted in the images using, for example, the Harris corner detector [35]. These corner points are easier to differentiate and a descriptor is estimated characterizing their location. Therefore, in many cases, feature-based methods are usually combined with optical flow providing higher accuracy and other measurements such as camera velocity.

In the first case, assuming that the differences between the camera viewpoints are small, the descriptors and the corresponding areas will be similar. Therefore, simple rectangular windows can be selected as the features centred on these points and methods based on cross-correlation, phase correlation or block matching can be utilized to estimate the transformation between them. These motion estimation methods over the images can be improved by reducing the error in image coordinate measurement, since it affects the accuracy of the obtained depth map [11, 36–38].

As it is expected, in cases of large changes in viewpoint's location between images, substantial differences are expected including changes in scale, occlusions and locations. Therefore, the task of automatic estimation of the correspondences is harder, but significant work has been performed in this area trying to provide solutions in these problems. Methods that provide rotation and scale-invariant feature registration [39–41], or that can handle affine transformations [42, 43] were proposed improving the accuracy of matching when the camera baseline is wide.

6.5   Epipolar Geometry

In order to obtain the structure of a scene from motion, epipolar geometry that is related to stereo vision is an essential part. It is assumed that two cameras view the same scene from two different viewpoints or that a single camera is moving capturing the same static scene from the same two locations. The projected 3D scene onto the 2D images generates relationships that can be described with matrices, allowing the estimation of the cameras' positions and the 3D locations of the scene points.

In Figure 6.5, an example of epipolar geometric is shown, indicating the two cameras observing the scene from their centres of projection. At this stage, few definitions

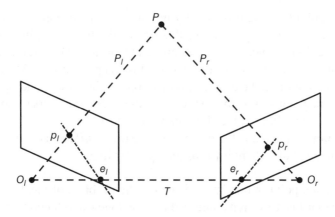

Figure 6.5 An example of epipolar geometric showing the two cameras observing the same scene from their centers of projection

will be provided that will be helpful to analyse the concepts related to scene reconstruction based on epipolar geometry.

- **Epipole** is the location of a camera's centre of projection projected into the other camera's image plane (see Figure 6.6). These two image points are also denoted by e_l and e_r in Figure 6.5, and with both centres of projection O_l and O_r lie on a single 3D line.
- **Epipolar lines** are the lines $(e_r p_r)$ and $(e_l p_l)$ in the right and left cameras, respectively. Observing Figure 6.5, it can be seen that the line $O_l P$ is seen by the left

Figure 6.6 An example showing the epipole projected into the other camera's image plane

camera as a point because it is directly in line with that camera's centre of projection. However, the right camera sees this line as the line $(e_r p_r)$ in its image plane, which is the epipolar line. These lines are related to the 3D location of point P and any line that intersects with the epipolar point is an epipolar line. Also, since the 3D line $O_l P$ passes through the centre of projection O_l, the corresponding epipolar line in the right image must pass through the epipole e_r and vice versa, showing that the points O_l, e_l, e_r and O_r are placed on the same line T.

- **Epipolar plane** is the plane formed by the two lines $O_l P$ and $O_r P$ intersecting each camera's image plane where it forms the epipolar lines.

Assuming that a point in one image is given, the problem of identifying the corresponding epipolar line to search along in the second view and 'encoding', the epipolar geometry can be approached using the essential and fundamental matrices. Let the essential matrix E relate the corresponding image points in the two pinhole camera models to projection matrices D_l and D_r [6]. Given a point P_r in the coordinate system of camera D_r, its position P_l in the coordinate of camera D_l is provided based on (6.1) by

$$P_r = R(P_l - T) \Rightarrow P_r^T E P_l = 0 \tag{6.21}$$

where R and T are the 3×3 rotation matrix and the translation 3×1 vector, respectively. S corresponds to the matrix representation of the cross product with T and is given by

$$T = \begin{bmatrix} 0 & -T_z & T_y \\ T_z & 0 & -T_x \\ -T_y & T_x & 0 \end{bmatrix} \tag{6.22}$$

And the essential matrix is equal to

$$E = RS \tag{6.23}$$

which depends only on the extrinsic parameters. Given that $p_l = (f_l/Z_l)P_l$ and $p_r = (f_r/Z_r)P_r$, Equation (6.21) is also true for the image points p_r and p_l, resulting the Longuet–Higgins equation.

$$P_r^T E P_l = 0 \Rightarrow ((Z_r/f_r)p_r)^T E((Z_l/f_l)p_l) = 0 \Rightarrow p_r^T E p_l = 0 \tag{6.24}$$

From the above, it can be observed that Equation (6.21) relates viewing rays, while Equation (6.24) relates 2D image points. This is expected since an image point can also be considered as a viewing ray. For example, a 2D image point (u,v) corresponds to a 3D point (u,v,f) on the image plane with the viewing ray into the scene been equal to $(fu/Z,fv/Z,f^2/Z)$ or to $(fX/Z,fY/Z,f)$ if the ray is passing through the point P.

Moving one step further, a considered line in the image is given by

$$au + bv + c = 0 \tag{6.25}$$

and that in a matrix form can be expressed as

$$\tilde{p}^T \tilde{l} = \tilde{l}^T \tilde{p} = 0 \tag{6.26}$$

where $\tilde{p} = \begin{bmatrix} u & v & 1 \end{bmatrix}^T$ and $\tilde{l} = \begin{bmatrix} a & b & c \end{bmatrix}^T$. Combining Equations (6.24) and (6.26), we can have that p_r belongs to the epipolar line in the right image defined by

$$\tilde{l}_r = E p_l \tag{6.27}$$

In the same way, p_l belongs to the epipolar line in the left image given by

$$\tilde{l}_l = E^T p_r \tag{6.28}$$

Since epipoles belong to the epipolar lines replacing them on Equations (6.27) and (6.28), we obtain

$$e_r^T E p_l = 0 \quad \text{and} \quad p_r^T E e_l = 0 \tag{6.29}$$

Also they belong to all the epipolar lines given by

$$e_r^T E = 0 \quad \text{and} \quad E e_l = 0 \tag{6.30}$$

which can be used to estimate the locations of the epipoles.

From Equation (6.23), the essential matrix is based on camera coordinates, and therefore, in order to use image coordinates, the intrinsic camera parameters need to be considered. From (6.13), inverting the camera calibration matrix C^{-1}, we obtain for the above points p_l and p_r that are in camera coordinates, their corresponding pixel coordinates.

$$\tilde{p}_l = C_l p_l \Rightarrow p_l = C_l^{-1} \tilde{p}_l \tag{6.31}$$

$$\tilde{p}_r = C_r p_r \Rightarrow p_r = C_r^{-1} \tilde{p}_r \tag{6.32}$$

Replacing Equations (6.31) and (6.32) in the Longuet–Higgins equation in (6.24), we obtain

$$(C_r^{-1} \tilde{p}_r)^T E (C_l^{-1} \tilde{p}_l) = 0 \Rightarrow \tilde{p}_r^T (C_r^{-1} E C_l^{-1}) \tilde{p}_l = 0 \Rightarrow \tilde{p}_r^T F \tilde{p}_l = 0 \tag{6.33}$$

where F is the fundamental matrix, with $F = C_r^{-1} R S C_l^{-1}$ depending both on intrinsic and extrinsic parameters (e.g. R, T, f). The fundamental matrix is a 3×3 matrix that has rank 2, and it can be estimated linearly given at least eight corresponding points. Significant work has been conducted in this area, focusing mainly on noisy images trying to accurately estimate the fundamental matrix [44–46].

The estimation of the fundamental matrix is essential to obtain the epipolar geometry. Assuming that pairs of point correspondences between the images are provided, a linear constrain on F is provided by Equation (6.33) for each of the point pairs. Based on the work of Hartley in [44] and assuming that some corresponding points have been estimated, we obtain a linear constraint on the fundamental matrix. Let $(x, y, 1)^T$ and $(x', y', 1)^T$ be the coordinates of a pair of corresponding points, the linear constrain for F will be

$$\begin{bmatrix} x' & y' & 1 \end{bmatrix} \begin{bmatrix} f_{11} & f_{12} & f_{13} \\ f_{21} & f_{22} & f_{23} \\ f_{31} & f_{32} & f_{33} \end{bmatrix} \begin{bmatrix} x \\ y \\ 1 \end{bmatrix} = 0 \quad or$$

$$xx' f_{11} + xy' f_{12} + xf_{13} + yx' f_{21} + yy' f_{22} + yf_{23} + x' f_{31} + y' f_{32} + f_{33} = 0 \quad (6.34)$$

Considering n pairs of points, we obtain a linear system of 9 unknown parameters in the matrix form that is written as

$$Hf = 0 \tag{6.35}$$

where H is an $n \times 9$ matrix with the estimated pairs of points, and f is a 9×1 vector containing the parameters of the fundamental matrix. To solve this system, at least 8 points are required and a least squares solution is utilized

$$\min_{\|f\|=1} \| Hf \|^2 \tag{6.36}$$

and this can be solved as the unit eigenvector with the minimum eigenvalue of $H^T H$. Also, it can be obtained numerically by using SVD. Reconstructing will provide the expected unique solution unless all the points lie on a plane [47].

6.6   Projection Matrix Recovery

The camera motion can be obtained using the camera calibration matrices C_r and C_l that consist of the rotation matrix R and the translation vector T. Therefore, using the calibration matrices, the camera motion in a 3D world coordinates can be calculated using the fundamental matrix that is obtained from the corresponding pairs of points as it was analysed earlier [48]. Based on the relation between the fundamental matrix and the essential matrix, we derive from Equation (6.33) that

$$E = C_l^T FC_r \tag{6.37}$$

and since the essential matrix is estimated, the camera matrix corresponding to the rotation between the viewpoints and the translation can be obtained based on Equation (6.23) and using SVD

$$E = U\Sigma V^T \tag{6.38}$$

where Σ is the diagonal matrix of singular values and U and V are orthogonal. The rotation matrix will then be given by

$$R = U \begin{bmatrix} 0 & -1 & 0 \\ 1 & 0 & 0 \\ 0 & 0 & 1 \end{bmatrix} V^T \tag{6.39}$$

and the translation by

$$S = U \begin{bmatrix} 0 & 1 & 0 \\ -1 & 0 & 0 \\ 0 & 0 & 0 \end{bmatrix} U^T \qquad (6.40)$$

The projection matrix D can be estimated from Equations (6.13, 6.39 and 6.40) as $D = C\begin{bmatrix} R & T \end{bmatrix}$.

6.6.1   Triangulation

Creating the 3D model of the captured scene requires also the triangulation of the 3D points. Methods such as [33, 48–52] provide optimal triangulations reconstructing the observed areas. Also, the Delaunay triangulation approach provides a simple solution to this problem mainly for short image sequences trying to connect 3D points that have a common edge.

6.7   Feature Detection and Registration

Reconstructing a 3D scene from multiple images requires the estimation of motion, which is based on feature detection in the most cases. This is due to the fact that global methods are not accurate enough in cases of occlusions or non-uniform lighting. Following a similar approach applied by humans, features are detected in the acquired images, and then pairs of points and their correspondences are estimated in order to estimate the motion and the related image transformations. The features should be selected in such a way that they will be coherent with the corresponding feature points in the reference image without being affected from geometric transformations (e.g. rotation, scaling) due to camera motion. Therefore, the most appropriate features will have the repeatability property, allowing them to be detected from all the viewpoints. In a scene and consequently in the captured images, corners, edges and blobs provide the most reliable features. The most well-known methods for feature detection in images are, for example, Harris, SIFT, Shi-Tomasi and FAST [53, 54].

Identifying locations that could provide good features is an important task of the feature detection process, and usually areas with gradients and significant variations in intensity are selected (see Figure 6.7). On the other hand, areas without any texture can match everywhere and therefore are not preferred for this task. In the case where the camera motion is known, the problem is simpler and edges are considered enough to detect the correspondences. In the cases where the camera motion is unknown, edges are not appropriate features due to the aperture problem. Therefore, the ideal features would be corners since they contain gradients with at least two orientations. Different methods have been designed for detecting corners, and the most common methods are analysed later.

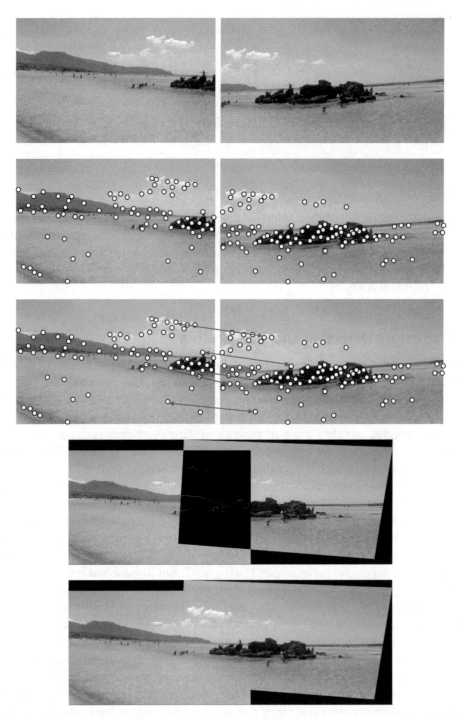

Figure 6.7 An example of the feature detection and extraction process, and the obtained aligned images (panoramic view)

6.7.1 Auto-correlation

Let us consider one image $I(x,y)$ and a second similar one, displaced by (u, v), noted as $I(x + u, y + v)$. Calculating the correlation between these two images, we obtain

$$C_{uv}(x, y) = \sum_{i=-N}^{N} \sum_{j=-N}^{N} I(x+i, y+j)I(x+u+i, y+v+j) \tag{6.41}$$

where $2N + 1$ is the size of the selected square block.

In case we consider small blocks over uniform areas, edges and corners (see figure 6.8), the obtained auto-correlation surfaces will have different shapes (e.g. flat, ridge-like or with a peak, respectively). So the steps to detect corners would be first to apply auto-correlation and then locate the coordinates with local maxima in both directions. This is essential because it distinguishes a corner from an edge that has only local maxima in either direction. Finally, a threshold is applied to consider that area as a corner or not. The correlation can be performed also in the frequency domain, or the sum of squared differences (SSD) could be used to reducing the computational complexity of this task.

$$SSD(x, y) = \sum_{i=-N}^{N} \sum_{j=-N}^{N} [I(x+u+i, y+v+j) - I(x+i, y+j)]^2 \tag{6.42}$$

6.7.2 Harris Detector

The Harris corner detector was proposed in [35] and is based on pixel gradients. In more detail, first-order partial derivatives are used to estimate the corner location on an image; and based on this concept similar implementations were suggested [55]. Let us assume $I(x,y)$ represents the intensity of an image, and using a Taylor expansion, we can approximate the pixel difference between the locations $(x + u, y + v)$

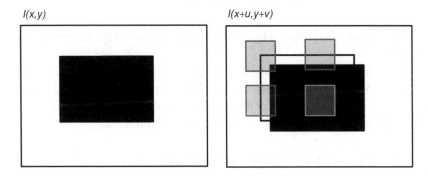

Figure 6.8 Example of block based auto-correlation over uniform areas, edges and corners

and (x, y) as

$$I(x + u, y + v) - I(x, y) = u\frac{\partial I(x, y)}{\partial x} + v\frac{\partial I(x, y)}{\partial y} \tag{6.43}$$

Replacing the intensity difference in (6.42) with Equation (6.43), we obtain

$$C(x, y) = \sum_i \sum_j \left(u\frac{\partial I(x, y)}{\partial x} + v\frac{\partial I(x, y)}{\partial y} \right)^2$$

$$= \sum_i \sum_j \left(u^2 \left(\frac{\partial I(x, y)}{\partial x} \right)^2 + 2uv\frac{\partial I(x, y)}{\partial x}\frac{\partial I(x, y)}{\partial y} + v^2 \left(\frac{\partial I(x, y)}{\partial y} \right)^2 \right) \tag{6.44}$$

where $\partial I/\partial x$ and $\partial I/\partial y$ are calculated at the location $(x + i, y + j)$. Removing the double summation and replacing a convolution with a matrix of ones J of size $(2N + 1) \times (2N + 1)$, we obtain

$$C(x, y) = u^2 \left(J * \left(\frac{\partial I(x, y)}{\partial x} \right)^2 \right) + 2uv \left(J * \frac{\partial I(x, y)}{\partial x}\frac{\partial I(x, y)}{\partial y} \right) + v^2 \left(J * \left(\frac{\partial I(x, y)}{\partial y} \right)^2 \right)$$

$$= (u \ v) \begin{pmatrix} a_{11} & a_{12} \\ a_{21} & a_{22} \end{pmatrix} \begin{pmatrix} u \\ v \end{pmatrix}$$

$$= A^T S A \tag{6.45}$$

where

$$a_{11} = J * \left(\frac{\partial I(x, y)}{\partial x} \right)^2$$

$$a_{12} = a_{21} = J * \frac{\partial I(x, y)}{\partial x}\frac{\partial I(x, y)}{\partial y}$$

$$a_{22} = J * \left(\frac{\partial I(x, y)}{\partial y} \right)^2 \tag{6.46}$$

The matrix J can also be replaced by a 2D Gaussian matrix of the same size smoothing the outcome.

A corner is characterized by a large variation of C in all directions, and obtaining the eigenvalues of S two 'large' eigenvalues should be expected. Based on the magnitudes of the eigenvalues (i.e. λ_1 and λ_2 for the large and small eigenvalues of S), the principal curvatures present in each considered block can be defined. So

- if $\lambda_1 \approx 0$ and $\lambda_2 \approx 0$, then there are no features present;
- if $\lambda_1 \approx 0$ and $\lambda_2 \gg 0$ (or vice versa), then an edge is present;
- if $\lambda_1 \gg 0$ and $\lambda_2 \gg 0$, then a corner is present.

This approach is computationally expensive as it computes the eigenvalue decomposition of S; therefore, Harris and Stephens suggested the following response function:

$$M = \lambda_1 \lambda_2 - \alpha(\lambda_1 + \lambda_2)^2 = \text{Det}(S) - \alpha(\text{Trace}(S))^2 \qquad (6.47)$$

where M is the response function that indicates the presence of a corner if it returns a large positive value, α is a tunable sensitivity parameter, $\text{Det}(S)$ is the determinant of matrix S, and $\text{Trace}(S)$ is the trace of matrix S. Based on this approach, Equation (6.47) does not require to compute the eigenvalues.

In order to increase the accuracy of registration and consequently the obtained SfM, the feature detector should be capable of localizing the same point on an object and be unaffected by any variation in scale, illumination, perspective imaging or partial occlusion. Harris corner detector has rotation invariance, partial invariance to additive and multiplicative intensity changes, but it is not invariant to image scale [56].

To overcome this problem, Lowe [57] proposed a model called scale-invariant feature transform (SIFT) that overcomes the issues due to scale variations. According to this method, image contents are transformed into local features that are invariant to translation rotation, scaling and other parameters (see Figure 6.9). These features are local and therefore robust to occlusions, they are distinctive since they can be matched to a large database of objects, and many of these features can be generated even for small objects. Also, the performance is close to real time, indicating the efficiency of this algorithm.

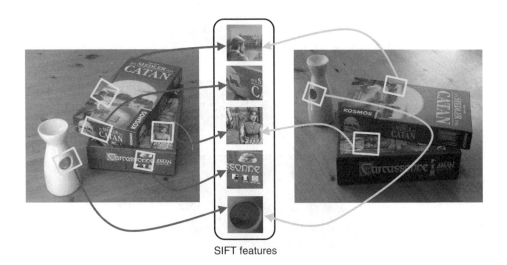

SIFT features

Figure 6.9   Examples of extracted SIFT features

6.7.3   SIFT Feature Detector

A brief analysis of the SIFT feature detector is presented. SIFT comprises four main stages. In the first stage, stable features across multiple scales are estimated using a Gaussian function of scale σ. The aim of this step is to identify locations and scales that can be re-estimated under different views of the same object. The outcome of this stage is the scale-space $L(x, y, \sigma) = G(x, y, \sigma) * I(x, y)$ of an image I, which is obtained by the convolution of a Gaussian multiscale kernel $G(x, y; \sigma) = \frac{1}{2\pi\sigma}\exp^{-(x^2+y^2)/2\sigma}$ with the given image (see figure 6.10).

Equivalently, the scale-space of an image can be defined as the solution of the diffusion equation (e.g. heat equation),

$$\partial_t L = \frac{1}{2}\nabla^2 L \tag{6.48}$$

with initial condition $L(x, y; 0) = I(x, y)$. The Laplacian of the Gaussian operator $\nabla^2 L(x, y, t)$ can also be computed as the difference between two Gaussian filtered (scale-space) images.

$$\nabla^2 L(x, y; \sigma) \approx \frac{\sigma}{\Delta\sigma}(L(x, y; \sigma + \Delta\sigma) - L(x, y; \sigma - \Delta\sigma)) \tag{6.49}$$

This approach is referred to as difference of Gaussians (DoG) and provides significant efficiency as it needs only to subtract images. Therefore, fast approximations are used based on box filters (by Integral Image). In order to create an Integral Image, first the Summed Area Table is calculated (see Figure 6.11). In this figure, the value

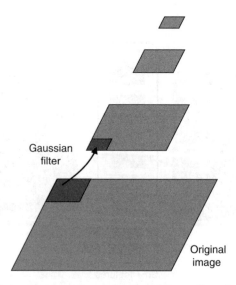

Gaussian
filter

Original
image

Figure 6.10   The pixels in a level are the result of applying a Gaussian filter to the lower level and then subsampling the reduce the size

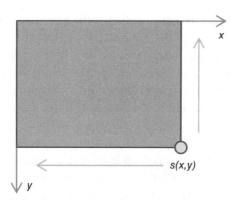

Figure 6.11   Example of calculating the Summed Area Table

at any point (x, y) is the sum of all the above-mentioned pixel values, to the left, and including the original pixel value of (x, y) itself.

$$s(x, y) = I(x, y) + s(x - 1, y) + s(x, y - 1) - s(x - 1, y - 1) \qquad (6.50)$$

Using the integral image representation, the value of any rectangular sum can be computed (see Figure 6.12). For example, rectangle D is equal to

$$I(x', y') = s(A) + s(D) - s(B) - s(C) \qquad (6.51)$$

Then the maxima and minima (extrema) of the DoG in scale-space are detected (location and scale), with each point compared to its eight neighbours in the current level and all the nine neighbours in the scales above and below.

In the second step, since the scale-space (Gaussian) and the corresponding DoG pyramids are generated, the locations of key points are estimated. Even the number of the obtained points is less than the image pixels; the amount of point is still significantly high to process them, and additionally only pixel accuracy is available.

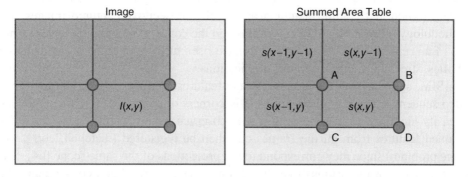

Figure 6.12   An example of an integral image representation

Therefore, in order to obtain the true location of the extrema, a Taylor series expansion is considered and the location of the minimum is estimated [57, 58]. The approach to reduce the number of points is based on removing all the locations with low contrast (low DoG). Also, points with strong edge responses in one direction only are removed following the same approach used in Harris corner detector. This step concludes the part of the process that is related to the detector. In the third step, SIFT descriptors are generated for each extrema. Initially, a region around each point is selected, and the effects of scale and rotation are removed. Since the correct image is chosen based on the point scale, the gradient magnitude m and orientation θ using finite differences are computed. Then a gradient histogram weighted by the magnitude and the Gaussian window is created. Finally, a parabola is fitted to the three histogram values closest to the peaks in order to interpolate them and obtain higher accuracy. In the final step, since the descriptor based on histogram orientations is created, the Lowe's second-nearest-neighbour algorithm is used to perform the matching process. Regarding the descriptor, it is a 128-dimensional feature vector obtained from a histogram of 4×4 samples per window in 8 directions. SIFT is one of the state-of-the-art feature detection algorithms. However, the speeded up robust features (SURF) [59] feature detection algorithm is much faster than SIFT, and the quality of feature detection and description is similar.

6.8   Reconstruction of 3D Structure and Motion

As it was discussed earlier in this chapter, SfM infers the 3D structure of a captured scene and the camera location over time, utilizing only sets of image correspondences. The state-of-the-art approaches such as the work presented in [60, 61] have commercial applications and are utilized in many everyday scenarios. The research conducted in this area tries to automate all the steps involved in this process, providing solutions to the problem of estimating the motion of the selected features in both cases of calibrated and uncalibrated cameras [48]. Furthermore, the feature selection process has been improved and unsupervised approaches have been proposed [57, 62]. The outcome of this process is to estimate the geometric properties of a scene based on registration and feature matching among the consecutive captured images. The methodology followed by SfM algorithms and the concept of obtaining the 3D structure of a scene and the camera position have been utilized in many areas including robotics, street view and panoramic applications.

In SfM, summarizing the process, image features are first extracted from a given video sequence. These features are usually corners or edges [35, 63] since they can be easily identified in a scene even if it is observed from different viewpoints. The obtained features from all the frames can then be registered (matched, correspondence problem) since they correspond to the projections of the same scene [64]. SfM in most of the cases assumes that the captured scene is rigid and consequently the observed motion is the relative motion between the viewer and the scene. This motion is estimated from the registered image features, and the epipolar geometry is calculated [6, 65]. The registration process can be improved by imposing further geometric constraints, and finally, based on the correspondences, triangulation is introduced to obtain the 3D structure of the observed scene [66].

6.8.1   Simultaneous Localization and Mapping

In the last few decades, there was significant progress in autonomous vehicle and robot navigation with the SLAM algorithm being the standard approach used to construct the environment on the fly and to estimate the ego-motion of the robotic system. SLAM is using mainly cameras as acquisition devices, but other sensors such as range/infrared cameras, laser devices and sonars can be considered. In more detail, SLAM in robotics and computer vision [67, 68] tries to generate map of an autonomous mobile robot's surroundings and to estimate its motion combining the data captured from one or multiple sensors. The work presented in [69, 70] provides solutions based on lasers; in [71, 72], radars and sonars were utilized, respectively; and in many cases, odometers were also included [73].

Monocular SLAM refers to the use of a monocular camera sensor [74], and considering the differences between SfM and monocular SLAM, the first approaches the problem without any restrictions on the acquisition devices, while monocular SLAM is based only on video sequences from a single camera. Furthermore, SLAM provides a sequential estimation and reconstruction of the surrounding area, which is essential for robotic and other applications such as augmented reality [75].

In monocular SLAM, information regarding the observed scene including camera parameters and detected features is transferred over time, and nowadays two main approaches are considered, key-frame and filter-based techniques. In the first case, two processes are performed simultaneously one operating in a high frame rate, while the second in a lower rate. During the first process on camera pose, estimation and tracking based on already detected features on the 3D scene is performed. The second process performs a global optimization trying to identify key-frames in the video sequence. In this approach, the second process is initialized from the first process [76–80], and the concept of key-frames is shown in Figure 6.13.

The second case tries to filter past information about the camera location and features that are not expected to be measured in the future, while it provides an estimate of the current position and the corresponding features on the 3D map based on joint probabilities. The most common algorithm for filtering using Gaussian probability distributions is the extended Kalman filter (EKF), and in Figure 6.14, its concept is illustrated showing that the marginalization of a past pose introduces probabilistic links between every pair of features in the map.

It is worth mentioning the work of Jung and Lacroix [81] on SLAM for stereo vision to perform terrain mapping. Also, Kim and Sukkarieh [82] used monocular vision in combination with inertial sensing to navigate UAVs to specified ground targets. The

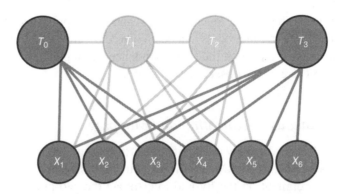

Figure 6.13   Visualisation of how inference progressed in a keyframe-based optimisation

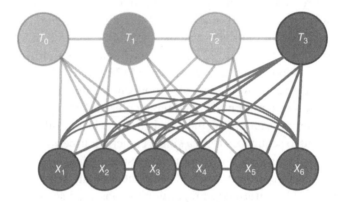

Figure 6.14   Visualisation of how inference progressed in a filter based optimisation

ATLAS mapping framework proposed by Bosse *et al.* [83] is based on omnidirectional vision, making particular use of lines. An algorithm that combines SIFT features [84] and FastSLAM filtering [85] was proposed by Sim *et al.* [86], and also it was considered by the commercial vSLAM system [87]. It can be observed that SIFT features provide high level of performance in registration process with a small trade-off in speed and computational complexity [74].

6.8.2   Registration for Panoramic View

Panoramic photography has significant and numerous applications, including image and video enhancement, compression and visualization. Also, commercial applications such as Google Street View and Bing Maps try to map and visualize the world's cities based on registration and other similar approaches. In these applications, the concept is to provide the tools to the users to navigate through streets and sights virtually by presenting a panoramic view captured at street level from a moving vehicle

[88, 89]. In order to synthesize a full panoramic view, several images are captured covering the whole viewing space. Then, these images are aligned trying to merge them into a complete panoramic image, and this process is called image stitching. The alignment part as it is also analysed in Chapter 4 is based on image registration, which is an essential step in panoramic photography. An example of image stitching that generates a panoramic view of 360 degrees is shown in Figure 6.15.

Since registration is one of the most important steps of this process many different approaches were proposed, incorporating either feature-based or area-based techniques. As it was previously mentioned, feature-based methods estimate image features (control points), such as corners and lines that can be registered between the given images. Since the correspondence problem is solved for a number of control points, a transformation matrix is computed providing the geometric relationship between the observed images. The next step of this approach is to apply this transformation in order to align the images. Regarding the area-based approaches, methods such as correlation (in the spatial or frequency domain), and template or block matching that are analysed in other chapters of this book, can be considered to estimate the correspondences between image regions. These methods can be applied globally for the whole image or locally, assuming that the given images are divided into overlapped or non-overlapped square regions. The following steps are similar to the feature-based approaches involving the estimation of the optimal geometric transformation and the alignment process. In Figure 6.16, the steps of a general image stitching algorithm are illustrated.

Recently, algorithms for image stitching based on the scale-invariance feature transform (i.e. SIFT) were proposed in [90, 91], providing reliable feature searching and matching solutions. The panoramic image mosaic method introduced by Szeliski [92] supports both local and global registration mechanisms to reduce the alignment errors. Methods that try to provide a robust feature-based image mosaic were proposed using RANSAC algorithm or SVD decomposition to eliminate outliers and obtain a more accurate transformation matrix between the images [93, 94]. An approach based on correlation was suggested in [95] with a Sobel edge detection filter to be used for

Figure 6.15 An example of image stitching that generates a panoramic view of 360 degrees, with the subjects moving around following the camera direction

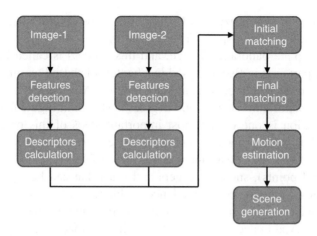

Figure 6.16   The main steps of a general image stitching algorithm

the extraction of features combined with a hierarchical matching mechanism. Also, methods for efficient stitching video sequences into panoramic mosaics have been proposed based on frame registration and on introducing the concept of key-frames (in a similar manner as it is used in SLAM) [96, 97].

6.9   Metrics and Datasets

The evaluation process of methods for SfM and all the related algorithms that are based on these concepts is usually performed separately on each submechanism of the overall approach. Since the overall performance depends on different aspects such as the accuracy of the feature selection and detection algorithms, they can be measured and compared separately. Also, evaluation frameworks have been proposed for the overall performance comparison in an application level.

Considering the evaluation of feature detectors for SfM algorithms, approaches based on the work of Martinez-Fonte in [98, 99] can be followed. In these cases, it is assumed that ground truth is available that is usually obtained from manual annotation. For example, in the case of the evaluation of corner detectors, ROC curves can be used to compare the performance of different methods. In this case, the probability of detection over the number of false alarms is plotted to illustrate the accuracy of each algorithm.

Another approach that can be utilized for the evaluation of SfM techniques and their expansions is based on the following steps. Initially, the calibration matrix that contains the intrinsic parameters of the camera used to capture testing images is obtained. The next step is to estimate the fundamental matrix, which expresses the relationship between any two given images. Then, using the fundamental matrices extracted from each algorithm under evaluation, the re-projection error generated by each feature detector is calculated [100]. The re-projection error (or geometric error) is the distance between a projected point (using the data extracted from the algorithms) and a measured point (ground-truth data). In order to obtain the ground truth, build-in devices such as odometry, laser scanners, depth cameras and GPS can be used, and in the case of multiple devices being present fusion can be applied. So, the rotation and translation vectors are extracted from the ground-truth data and then compared against the results obtained from each algorithm using the Euclidean distance between them after all the data were normalized. Furthermore, the fundamental matrix can be also assessed based on the structure of the corresponding epipolar geometry [101]. Therefore, it is not useful to compare two solutions by directly comparing the difference in their fundamental matrices. Thus, the differences in the associated epipolar geometries are compared, weighted by the density of the given matching points.

In the case of a SLAM system, the output is the estimated camera trajectory and the map of the observed area. Also, it is not always possible to obtain the ground truth of the reconstructed map, and therefore often data captured from depth cameras are utilized. In more detail, as it was analysed earlier, a sequence of poses from the estimated $P_1, \dots, P_n$ and the ground truth $G_1, \dots, G_n$ trajectory is assumed that are provided. Also, it is assumed that these sequences have the same length n, which are equally sampled and time-synchronized.

One of the metrics used is the relative pose error that calculates the local accuracy of the trajectory over a specified time duration Δt

$$\varepsilon_t = (G_t^{-1} G_{t+\Delta t})^{-1} (P_t^{-1} P_{t+\Delta t}) \tag{6.52}$$

From this error, the root-mean-squared error (RMSE) of the translational component averaged over all possible time intervals Δt is obtained as

$$\text{RMSE}(\varepsilon_{1:n}) = \frac{1}{n} \sum_{\Delta t}^{n} \left(\frac{1}{m} \sum_{t=1}^{m} \| \text{ trans}(\varepsilon_t) \|^2 \right)^{1/2} \tag{6.53}$$

where n is the number of camera poses, and $m = n - \Delta t$ is the obtained individual relative to pose errors. Also, the mean error is utilized instead of the RMSE, since in many cases it provides less influence to outliers [102–104].

Regarding the evaluation process based on the global consistency, the absolute distances between the estimated and the ground-truth trajectory are compared as it was discussed earlier. Before the comparative evaluation, the trajectories need to be aligned since they may be specified in arbitrary coordinate frames (e.g. GPS, odometry) [105]. After the alignment process is completed, a rigid-body transformation T is obtained, mapping the estimated $P_{1:n}$ onto the ground-truth $G_{1:n}$ trajectory. Consequently, the absolute trajectory error is given by

$$\tau_t = G_t^{-1} S P_t \tag{6.54}$$

Thus, the overall RMSE for the translational components is given by

$$\mathrm{RMSE}(\tau_{1:n}) = \left(\frac{1}{n} \sum_{t}^{n} \| \ \mathrm{trans}(\tau_t) \ \|^2 \right)^{1/2} \tag{6.55}$$

where t is the time step.

Finally, the evaluation process could include comparison of the feature detection methods, indicating that approaches with fewer features could provide similar performance and accuracy with others that produce more but less accurate ones. Consequently, the overall complexity and the time requirements to complete the reconstruction process are essential parameters to evaluate and compare different methods. Therefore, time, complexity and frame rate comparisons are utilized to provide a comparative study of SfM algorithms, their subcomponents and related applications, where SfM is a component of the whole system.

6.9.1   Datasets for Performance Evaluation

Datasets are essential for the evaluation of the SfM techniques, and systems for SLAM or panoramic view. Several databases with images and the related data captured in different modalities are available, including ground truth and suggested metrics. The 'Photo tourism' dataset [106, 107] provides images of outdoor buildings and monuments, aiming to reconstruct large-scale areas even cities. Another dataset for reconstruction of buildings in urban environments is available in [108] providing the input images and the per-image camera projection matrices. In [109], a dataset for SfM is available providing also the tracked image points and the final reconstructions. An indoor dataset captured using Kinect and another one with a car are available in [110], focusing on semantic SfM scene analysis. In Pittsburgh fast-food image dataset [111], a large set of images of food items is provided captured from different directions. A dataset for image stitching is available in [112], providing a test suite of 22 challenging image pairs. In [113], a dataset with cars captured from different points of view is available. A dataset for non-rigid SfM is provided in [114], while data for large-scale SfM are available in [115, 116]. Finally, the 'BigSFM: Reconstructing the World from Internet Photos' database provides RGB-D data and ground-truth data for the evaluation of visual odometry and visual SLAM systems. This dataset contains the colour and depth images of a Microsoft Kinect sensor along with its ground-truth trajectory. The data are recorded at full frame rate (30 Hz) and sensor resolution (640 × 480). A high-accuracy motion capture system was used to obtain the ground-truth trajectory and also the accelerometer data from the Kinect are provided [117].

6.10   Conclusion

In this chapter, the registration problem was discussed for applications related to SfM, SLAM and image stitching. The most common approaches used to solve the matching problem in these areas based on feature detection were introduced. Initially, an overview of the problem and the concept of SfM were presented, and the steps to estimate the structure of the observed scene were further analysed. An overview of the pinhole camera model was provided followed by theoretical concepts and mechanisms to perform camera calibration. The correspondence problem and the related solutions were discussed, and details of the conclusion of the main steps of SfM method on epipolar geometry and triangulation were provided.

The part of this chapter focused on feature detection and matching that is related to registration was further analysed, providing details for the Harris and the SIFT feature detectors. In the following sections, systems for reconstruction of 3D structures and mapping using SLAM were presented, and also an overview of the steps applied in image stitching algorithms was available. Finally, metrics and datasets used to evaluate SfM methods, SLAM systems and panoramic view techniques were discussed indicating the scale of the applications, and the level of details and accuracy of the state-of-the-art approaches.

6.11   Exercise–Practice

This exercise is focused on developing an image stitching algorithm for panoramic view applications. Following the steps in Figure 6.16 align the images provided in Figure 6.7. As it was mentioned feature-based methods estimate image features, such as corners that can be registered between the given images. In this stage, the corresponding features can be manually selected. Since the correspondence problem is solved for a number of control points, calculate the transformation matrix to obtain the geometric relationship between the observed images. Then apply this transformation in order to align the images.

Finally, in order to automate the process the features and their correspondences can be estimated automatically using one of the discussed algorithms. Free Matlab implementations of these feature selection and registration algorithms are available on the Web.

References

[1] Dellaert, F., Seitz, S., Thorpe, C. and Thrun, S. (2000) Structure from motion without correspondence. *Proceedings of the IEEE Computer Society Conference on Computer Vision and Pattern Recognition*, 2000, Vol. **2**, pp. 557–564.
[2] Tomasi, C. (1992) Shape and motion from image streams under orthography: a factorization method. *International Journal of Computer Vision*, **9**, 137–154.
[3] Poelman, C. and Kanade, T. (1997) A paraperspective factorization method for shape and motion recovery. *IEEE Transactions on Pattern Analysis and Machine Intelligence*, **19** (3), 206–218.

[4] Morris, D. and Kanade, T. (1998) A unified factorization algorithm for points, line segments and planes with uncertainty models. *Proceedings of 6th IEEE International Conference on Computer Vision (ICCV'98)*, pp. 696–702.

[5] Pollefeys, M., Koch, R. and Gool, L.V. (1999) Self-calibration and metric reconstruction in spite of varying and unknown internal camera parameters. *International Journal of Computer Vision*, pp. 7–25.

[6] Faugeras, O. (1993) *Three-Dimensional Computer Vision: A Geometric Viewpoint*, The MIT press, Cambridge, MA.

[7] Faugeras, O. and Keriven, R. (1998) Complete dense stereovision using level set methods. *Proceedings of the 5th European Conference on Computer Vision*, pp. 379–393.

[8] Kutulakos, K. and Seitz, S. (1999) A theory of shape by space carving. Proceedings of the 7th International Conference on Computer Vision, pp. 307–314.

[9] Huttenlocher, D. and Ullman, S. (1990) Recognizing solid objects by alignment with an image. *International Journal of Computer Vision*, **5** (2), 195–212.

[10] Lowe, D. (1991) Fitting parameterized three-dimensional models to images. *IEEE Transactions on Pattern Analysis and Machine Intelligence*, **13** (5), 441–450.

[11] Beardsley, P., Torr, P. and Zisserman, A. (1996) 3d model acquisition from extended image sequences. European Conference on Computer Vision (ECCV), Vol. 2, pp. 683–695.

[12] Zhang, Z. (1998) Determining the epipolar geometry and its uncertainty - a review. *International Journal of Computer Vision*, **27** (2), 161–195.

[13] Forsyth, D., Ioffe, S. and Haddon, J. (1999) Bayesian structure from motion. International Conference on Computer Vision (ICCV), pp. 660–665.

[14] Zhang, Z. (2004) Camera calibration, *Emerging Topics in Computer Vision*, Chapter 2, Prentice Hall Professional Technical Reference, pp. 4–43.

[15] Wei, G. and Ma, S. (1994) Implicit and explicit camera calibration: theory and experiments. *IEEE Transactions on Pattern Analysis and Machine Intelligence*, **16** (5), 469–480.

[16] Agrawal, M. and Davis, L. (2003) Camera calibration using spheres: a semi-definite programming approach. Proceedings of the 9th International Conference on Computer Vision, October 2003, IEEE Computer Society Press, pp. 782–789.

[17] Knorr, M., Niehsen, W. and Stiller, C. (2013) Online extrinsic multi-camera calibration using ground plane induced homographies. Intelligent Vehicles Symposium (IV), 2013 IEEE, June 2013, pp. 236–241.

[18] Rosebrock, D. and Wahl, F. (2012) Generic camera calibration and modeling using spline surfaces. Intelligent Vehicles Symposium (IV), 2012 IEEE, June 2012, pp. 51–56.

[19] Melo, R., Barreto, J. and Falcao, G. (2012) A new solution for camera calibration and real-time image distortion correction in medical endoscopy initial technical evaluation. *IEEE Transactions on Biomedical Engineering*, **59** (3), 634–644.

[20] Zhang, Z. (2004) Camera calibration with one-dimensional objects. *IEEE Transactions on Pattern Analysis and Machine Intelligence*, **26** (7), 892–899.

[21] Zhang, Z., Li, M., Huang, K. and Tan, T. (2008) Practical camera auto-calibration based on object appearance and motion for traffic scene visual surveillance. IEEE Conference on Computer Vision and Pattern Recognition, June 2008. CVPR 2008, pp. 1–8.

[22] Takahashi, K., Nobuhara, S. and Matsuyama, T. (2012) A new mirror-based extrinsic camera calibration using an orthogonality constraint. 2012 IEEE Conference on Computer Vision and Pattern Recognition (CVPR), June 2012, pp. 1051–1058.

[23] Zhang, Z. (2000) A flexible new technique for camera calibration. *IEEE Transactions on Pattern Analysis and Machine Intelligence*, **22** (11), 1330–1334.

[24] Robert, L. (1995) Camera Calibration Without Feature Extraction. Computer Vision, Graphics, and Image Processing, INRIA Technical Report 2204, Vol. 63, No. 2, pp. 314–325.

[25] Press, W., Teukolsky, S., Vetterling, W. and Flannery, B. (1992) *Numerical Recipes in C: The Art of Scientific Computing*, Cambridge University Press.

[26] Abdel-Aziz, Y. and Karara, H. (1971) Direct linear transformation into object space coordinates in close-range photogrammetry. Proceedings of Symposium on Close-Range Photogrammetry, Urbana, IL, pp. 1–18.

[27] Faugeras, O. and Toscani, G. (1987) Camera calibration for 3d computer vision. Proceedings of the International Workshop on Industrial Applications of Machine Vision and Machine Intelligence, pp. 240–247.

[28] Melen, T. (1994) Geometrical modelling and calibration of video cameras for underwater navigation. PhD thesis, Norges tekniske hogskole, Institutt for teknisk kybernetikk.

[29] Heikkila, J. and Silven, O. (1997) A four-step camera calibration procedure with implicit image correction. Proceedings, 1997 IEEE Computer Society Conference on Computer Vision and Pattern Recognition, June 1997, pp. 1106–1112.

[30] Debevec, P., Taylor, C. and Malik, J. (1996) Modelling and rendering architecture from photographs: a hybrid geometry- and image-based approach. Computer Graphics (SIGGRAPH'96).

[31] Fakih, A. and Zelek, J. (2008) Structure from motion: combining features correspondences and optical flow. 19th International Conference on Pattern Recognition, December 2008. ICPR 2008, pp. 1–4.

[32] Pollefeys, M., Gool, L., Vergauwen, M. *et al.* (2004) Visual modeling with a hand-held camera. *International Journal of Computer*, **59** (3), 207–232.

[33] Lhuillier, M. and Quan, L. (2005) A quasi-dense approach to surface reconstruction from uncalibrated images. *IEEE Transactions on Pattern Analysis and Machine Intelligence*, **27** (3), 418–433.

[34] Nister, D. (2005) Preemptive RANSAC for live structure and motion estimation. *Machine Vision and Applications*, **16** (5), 321–329.

[35] Harris, C. and Stephens, M. (1988) A combined corner and edge detector. Alvey Vision Conference, pp. 189–192.

[36] Kanade, T. and Okutomi, M. (1994) A stereo matching algorithm with an adaptative window: theory and experiment. *IEEE Transactions on Pattern Analysis and Machine Intelligence*, **16** (9), 920–932.

[37] Zisserman, A., Fitzgibbon, A. and Cross, G. (1999) VHS to VRML: 3d graphical models from video sequences. ICMCS, pp. 51–57.

[38] Pollefeys, M., Koch, R. and Gool, L. (1998) Self-calibration and metric reconstruction in spite of varying and unknown internal camera parameters. International Conference on Computer Vision, pp. 90–95.

[39] Schmid, C., Mohr, R. and Bauckhage, C. (1998) Comparing and evaluating interest points. International Conference on Computer Vision, pp. 230–235.

[40] Baumberg, A. (2000) Reliable feature matching across widely separated views. IEEE Computer Society Conference on Computer Vision and Pattern Recognition, pp. 774–781.

[41] Mikolajczyk, K. and Schmid, C. (2001) Indexing based on scale invariant interest points. International Conference on Computer Vision, pp. 525–531.

[42] Pritchett, P. and Zisserman, A. (1998) Wide baseline stereo matching. International Conference on Computer Vision, pp. 754–760.

[43] Tuytelaars, T. and Gool, L. (1999) Content-based image retrieval based on local affinely invariant regions. International Conference on Visual Information Systems, pp. 493–500.

[44] Hartley, R. (1995) In defence of the 8-point algorithm. International Conference on Computer Vision, pp. 1064–1070.

[45] Torr, P. and Zisserman, A. (1998) Robust computation and parameterization of multiple view relations. International Conference on Computer Vision, pp. 727–732.

[46] Zhang, Z., Deriche, R., Faugeras, O. and Luong, Q. (1995) A robust technique for matching two uncalibrated images through the recovery of the unknown epipolar geometry. *Artificial Intelligence Journal*, **78**, 87–119.

[47] Faugeras, O. and Maybank, S. (1990) Motion from point matches: multiplicity of solutions. *International Journal of Computer Vision*, **4** (3), 225–246.

[48] Hartley, R. and Zisserman, A. (2004) *Multiple View Geometry in Computer Vision*, Cambridge University Press.

[49] Birchfield, S. and Tomasi, C. (1999) Depth discontinuities by pixel-to-pixel stereo. *International Journal of Computer Vision*, **3**, 269–293.

[50] Cox, I., Hingorani, S., Rao, S. and Maggs, B. (1996) A maximum likelihood stereo algorithm. *Computer Vision and Image Understanding*, **63** (3), 542–567.

[51] Pollefeys, M., Gool, L., Meerbergen, G.V. and Vergauwen, M. (2002) A hierarchical symmetric stereo algorithm using dynamic programming. *International Journal of Computer Vision*, **47**, 275–285.

[52] Woodford, O., Torr, P., Reid, I. and Fitzgibbon, A. (2008) Global stereo reconstruction under second order smoothness priors. Proceedings of IEEE Conference on Computer Vision and Pattern Recognition.

[53] Castle, R. (2013) *Object Recognition for Wearable Visual Robots*, University of Oxford.

[54] Shi, J. and Tomasi, C. (1994) Good features to track. Computer Vision and Pattern Recognition, pp. 593–600.

[55] Xiao-Chen, Y. and Chi-Man, P. (2011) Invariant digital image watermarking using adaptive harris corner detector. 8th International Conference on Computer Graphics, Imaging and Visualization, pp. 109–113.

[56] MIT (2013) Advances in computer vision. Lecture 10, Invariant feature.

[57] Lowe, D. (2004) Distinctive image features from scale-invariant keypoints. *International Journal of Computer Vision*, **60** (2), 91–110.

[58] Brown, M. and Lowe, D. (2002) Invariant features from interest point groups. British Machine Vision Conference, pp. 656–665.

[59] Bay, H., Ess, A., Tuytelaars, T. and Van Gool, L. (2008) Speeded-up robust features (SURF). *Computer vision and Image Understanding*, **9** (110), 346–359.

[60] PhotoTourism (2011) Exploring Photo Collections in 3D, http://phototour.cs.washington.edu/ (accessed 13 February 2015).

[61] AutoSticht (2011) AutoStitch: A New Dimension in Automatic Image Stitching, http://www.autostitch.net/ (accessed 13 February 2015).

[62] Triggs, B., McLauchlan, P., Hartley, R. and Fitzgibbon, A. (2000) Bundle adjustment - a modern synthesis. Vision Algorithms: Theory and Practice, pp. 298–375.

[63] Canny, J. (1986) A computational approach to edge detection. *IEEE Transactions on Pattern Analysis and Machine Intelligence*, **PAMI-8** (6), 679–698.

[64] Marr, D. and Poggio, T. (1976) *Cooperative Computation of Stereo Disparity*, Massachusetts Institute of Technology.

[65] Baker, H. and Binford, T. (1981) Depth from edge and intensity based stereo. Proceedings of the 7th International Joint Conference on Artificial Intelligence, Vol. 2, IJCAI'81, pp. 631–636.

[66] Hartley, R. and Sturm, P. (1997) Triangulation. *Computer Vision and Image Understanding*, **68** (2), 146–157.

[67] Durrant-Whyte, H. and Bailey, T. (2006) Simultaneous localization and mapping: part I. *IEEE Robotics and Automation Magazine*, **13** (2), 99–110.

[68] Bailey, T. and Durrant-Whyte, H. (2006) Simultaneous localization and mapping (SLAM): part II. *IEEE Robotics and Automation Magazine*, **13** (3), 108–117.

[69] Castellanos, J., Montiel, J., Neira, J. and Tardos, J. (1999) The SPmap: a probabilistic framework for simultaneous localization and map building. *IEEE Transactions on Robotics and Automation*, **15** (5), 948–952.

[70] Newman, P., Leonard, J., Neira, J. and Tardos, J. (2002) Explore and return: experimental validation of real time concurrent mapping and localization. Proceedings of the IEEE International Conference on Robotics and Automation, pp. 1802–1809.

[71] Dissanayake, M., Newman, P., Clark, S. *et al.* (2001) A solution to the simultaneous localization and map building (SLAM) problem. *IEEE Transactions on Robotics and Automation*, **17** (3), 229–241.

[72] Paz, L., Tardos, J. and Neira, J. (2008) Divide and conquer: EKF SLAM in O(n). *IEEE Transactions on Robotics*, **24** (5), 1107–1120.

[73] Castellanos, J., Neira, J. and Tardos, J. (2001) Multisensor fusion for simultaneous localization and map building. *IEEE Transactions on Robotics and Automation*, **17** (6), 908–914.

[74] Davison, A., Reid, I., Molton, N. and Stasse, O. (2007) MonoSLAM: real-time single camera SLAM. *IEEE Transactions on Pattern Analysis and Machine Intelligence*, **29** (6), 1052–1067.

[75] Klein, G. and Murray, D. (2009) Parallel tracking and mapping on a camera phone. 8th IEEE International Symposium on Mixed and Augmented Reality, October 2009. ISMAR 2009, pp. 83–86.

[76] Lovegrove, S. and Davison, A. (2010) Real-time spherical mosaicing using whole image alignment. Proceedings of the 11th European Conference on Computer Vision Conference on Computer Vision: Part III, ECCV'10, pp. 73–86.

[77] Handa, A., Chli, M., Strasdat, H. and Davison, A. (2010) Scalable active matching. 2010 IEEE Conference on Computer Vision and Pattern Recognition (CVPR), June 2010, pp. 1546–1553.

[78] Konolige, K. and Agrawal, M. (2008) FrameSLAM: from bundle adjustment to real-time visual mapping. *IEEE Transactions on Robotics*, **24** (5), 1066–1077.

[79] Strasdat, H., Montiel, J. and Davison, A. (2010) Scale drift-aware large scale monocular SLAM. Proceedings of Robotics: Science and Systems (RSS).

[80] Konolige, K., Agrawal, M. and Solra, J. (2007) Large-scale visual odometry for rough terrain. Proceedings of the International Symposium on Research in Robotics, pp. 201–212.

[81] Jung, I.-K. and Lacroix, S. (2003) High resolution terrain mapping using low attitude aerial stereo imagery. Proceedings of the 19th IEEE International Conference on Computer Vision, October 2003, Vol. 2, pp. 946–951.

[82] Kim, J.-H. and Sukkarieh, S. (2003) Airborne simultaneous localisation and map building. IEEE International Conference on Robotics and Automation, September 2003. Proceedings. ICRA '03., Vol. 1, pp. 406–411.

[83] Bosse, M., Rikoski, R., Leonard, J. and Teller, S. (2002) Vanishing points and 3d lines from omnidirectional video. Proceedings 2002 International Conference on Image Processing, June 2002, Vol. 3, pp. 513–516.

[84] Lowe, D. (1999) Object recognition from local scale-invariant features. Proceedings of the IEEE Computer Society Conference on Computer Vision and Pattern Recognition, Vol. 2, pp. 1150–1157.

[85] Michael, M., Sebastian, T., Daphne, K. and Ben, W. (2002) FastSLAM: a factored solution to the simultaneous localization and mapping problem. 18th National Conference on Artificial Intelligence, pp. 593–598.

[86] Sim, R., Elinas, P., Griffin, M. and Little, J. (2005) Vision-based SLAM using the rao-blackwellised particle filter. Proceedings of the IJCAI Workshop on Reasoning with Uncertainty in Robotics.

[87] Karlsson, N., Bernardo, E., Ostrowski, J. *et al.* (2005) The vSLAM algorithm for robust localization and mapping. Proceedings of the 2005 IEEE International Conference on Robotics and Automation, April 2005. ICRA 2005, pp. 24–29.

[88] Sallal, K. and Rahma, A. (2013) Feature based registration for panoramic image generation. *IJCSI International Journal of Computer Science Issues*, **10** (6), 132–138.

[89] Zheng, E., Raguram, R., Fite-Georgel, P. and Frahm, J.-M. (2011) Efficient generation of multi-perspective panoramas. 2011 International Conference on 3D Imaging, Modeling, Processing, Visualization and Transmission (3DIMPVT), May 2011, pp. 86–92.

[90] Lin, L. and Nan, G. (2010) Algorithm for sequence image automatic mosaic based on sift feature. WASE International Conference on Information Engineering.

[91] Prajkta, S., Krishnan, K. and Anita, P. (2011) A novel approach for generation of panoramic view. *(IJCSIT) International Journal of Computer Science and Information Technologies*, **2** (2), 804–807.

[92] Szeliski, R. (1996) Video mosaics for virtual environments. IEEE Computer Graphics and Applications, pp. 22–30.

[93] Bing, W. and Xiaoli, W. (2013) Video serial images registration based on FBM algorithm. *Research Journal of Applied Sciences, Engineering and Technology*, **5** (17), 4274–4278.

[94] Subramanyam, M.V. (2013) Feature based image mosaic using steerable filters and harris corner detector. *I.J. Image, Graphics and Signal Processing*, **6**, 9–15.

[95] Asha, R., Asha, K. and Manjunath, M. (2013) Image mosaicing and registration. *International Journal of Computer Science Issues*, **10** (2), 534–540.

[96] Steedly, D., Pal, C. and Szeliski, R. (2005) Efficiently registering video into panoramic mosaics. 10th International Conference on Computer Vision (ICCV 2005), October 2005, IEEE Computer Society, pp. 1300–1307.

[97] Aseem, A., Colin, Z.K., Chris, P. *et al.* (2005) Panoramic video textures. ACM SIGGRAPH 2005 Papers, SIGGRAPH '05, pp. 821–827.

[98] Cooke, T. and Whatmough, R. (2005) Evaluation of corner detectors for structure from motion problems. DICTA, p. 77.

[99] Martinez-Fonte, L., Gautama, S. and Philips, W. (2004) An empirical study on corner detection to extract buildings in very high resolution satellite images. IEEE-ProRisc, pp. 288–293.

[100] Govender, N. (2009) Evaluation of feature detection algorithms for structure from motion. 3rd Robotics and Mechatronics Symposium (ROBMECH), Technical Report, Council for Scientific and Industrial Research, Pretoria.

[101] Torr, P.H.S. (2004) A Structure and Motion Toolkit in MATLAB: Interactive Adventures in S and M. Technical Report, MRS-TR-2002-5, Microsoft Research.

[102] Sturm, J., Burgard, W. and Cremers, D. (2012) Evaluating egomotion and structure-from-motion approaches using the tum RGB-D benchmark. Proceedings of the Workshop on Color-Depth Camera Fusion in Robotics at the IEEE/RJS International Conference on Intelligent Robot Systems (IROS), October 2012.

[103] Kelly, A. (2004) Linearized error propagation in odometry. *International Journal of Robotics Research (IJRR)*, **23** (2), 179–218.

[104] Kummerle, R., Steder, B., Dornhege, C. *et al.* (2009) On measuring the accuracy of SLAM algorithms. *Autonomous Robots*, **27**, 387–407.

[105] Horn, B. (1987) Closed-form solution of absolute orientation using unit quaternions. *Journal of the Optical Society of America A*, **4**, 629–642.

[106] Snavely, N., Seitz, S.M. and Szeliski, R. (2006) Photo tourism: exploring photo collections in 3d, *SIGGRAPH Conference Proceedings*, ACM Press, New York, pp. 835–846.

[107] Agarwal, S., Furukawa, Y., Snavely, N. *et al.* (2011) Building rome in a day. *Communications of the ACM*, **54** (10), 105–112.

[108] Ceylan, D., Mitra, N., Zheng, Y. and Pauly, M. (2014) Coupled structure-from-motion and 3d symmetry detection for urban facades. ACM Transactions on Graphics.

[109] Enqvist, O., Olsson, C. and Kahl, F. (2011) Stable Structure from Motion Using Rotational Consistency. Technical Report, Centre for Mathematical Sciences, Lund University.

[110] Bao, S.Y., Bagra, M., Chao, Y.-W. and Savarese, S. (2012) Semantic structure from motion with points, regions, and objects. Proceedings of the IEEE International Conference on Computer Vision and Pattern Recognition.

[111] Chen, M., Yang, J. and Dhingra, K.D. (2009) *Pittsburgh Fast-Food Image Dataset (PFID)*, Intel Labs Pittsburgh.

[112] Yang, G., Stewart, C., Sofka, M. and Tsai, C.-L. (2007) Registration of challenging image pairs: initialization, estimation, and decision. *IEEE Transactions on Pattern Analysis and Machine Intelligence*, **29** (11), 1973–1989.

[113] Ozuysal, M., Lepetit, V. and Fua, P. (2009) Pose estimation for category specific multiview object localization. Conference on Computer Vision and Pattern Recognition, Miami, FL, June 2009.

[114] Torresani, L., Hertzmann, A. and Bregler, C. (2008) Non-rigid structure-from-motion: estimating shape and motion with hierarchical priors. *IEEE Transactions on Pattern Analysis and Machine Intelligence*, **30** (5), 878–892.

[115] Crandall, D., Owens, A., Snavely, N. and Huttenlocher, D.P. (2011) Discrete-continuous optimization for large-scale structure from motion. Proceedings of the IEEE Conference on Computer Vision and Pattern Recognition.

[116] Crandall, D., Owens, A., Snavely, N. and Huttenlocher, D. (2012) SFM with MRFs: discrete-continuous optimization for large-scale reconstruction. IEEE Transactions on Pattern Analysis and Machine Intelligence.

[117] Sturm, J., Engelhard, N., Endres, F. *et al.* (2012) A benchmark for the evaluation of RGB-D SLAM systems. Proceedings of the International Conference on Intelligent Robot Systems (IROS), October 2012.

7

Medical Image Registration Measures

7.1 Introduction

Registration is a problem that underpins many tasks in medical image analysis. It may be required to compare images from a patient acquired at different times, to combine images acquired from different imaging systems or to put images from different subjects into a common coordinate system for atlas construction or cross-sectional studies. It is convenient to classify medical image registration problems according to whether the images to be aligned are from the same or different modalities (monomodal vs multimodal registration), and whether they represent the same or different subjects (intra-subject vs inter-subject registration).

Monomodal registration can usually be performed using classical registration measures used in other sub-fields of computer vision, such as video tracking, pattern recognition, stereomapping or remotely sensed data processing [1]. Multimodal registration, however, has necessitated the development by the medical image analysis community of new registration measures such as mutual information that are adapted to the situation where there exists a complex many-to-many relationship between image intensities of equivalent structures.

Another characteristic of most medical images is that they are three-dimensional in nature; therefore, image registration usually involves estimating a three-dimensional spatial transformation. In intra-subject registration, the transformation that relates the images is often rigid (consisting of a translation and a rotation), but may be non-rigid if tissue deformations or geometrical distortions inherent to the imaging processes need to be compensated for. Inter-subject registration always involves non-rigid transformations. For instance, brain image warping methods [2, 3] necessitate large deformation models that go well beyond traditional optical flow techniques in order to account for anatomical variations across individuals.

Image, Video & 3D Data Registration: Medical, Satellite & Video Processing Applications with Quality Metrics,
First Edition. Vasileios Argyriou, Jesus Martinez Del Rincon, Barbara Villarini and Alexis Roche.
© 2015 John Wiley & Sons, Ltd. Published 2015 by John Wiley & Sons, Ltd.

In this chapter, we shall focus on the specifics of medical image registration, which are also apparent in the various reviews of the topic written over the years [4–8]. Sections 7.2 and 7.3 expose a mathematical framework for the construction of registration measures adapted to medical image analysis problems, starting from a general feature-based registration perspective and then highlighting widely used intensity-featured measures as special cases. The remaining Section 7.4 presents several common three-dimensional transformation models ranging from rigid transformations to free-form large deformation models, and discusses some associated optimization procedures.

7.2   Feature-Based Registration

Feature-based registration methods use geometrical objects, such as points, oriented points, lines, surfaces, to drive the registration of two images. These methods therefore rely on a preliminary segmentation step in which homologous features are extracted from both images. Formally, assume that we wish to register a source image I with a floating image J from two sets of homologous features, $I = \{\mathbf{f}_1, \mathbf{f}_2, \ldots, \mathbf{f}_n\} \subset \mathcal{F}$ and $J = \{\mathbf{g}_1, \mathbf{g}_2, \ldots, \mathbf{g}_m\} \subset \mathcal{F}$, where $\mathcal{F}$ denotes the chosen feature space. The goal is then to find a spatial transformation $T : \mathbb{R}^3 \to \mathbb{R}^3$ in a given set of allowable transformations $\mathcal{T}$ that best aligns I with J in some sense.

Following [9], we may define a general registration algorithm that works with any feature type. Let us assume that both feature sets I and J are *partially homologous* in the sense that there exists a discrete mapping $Q : I \to J \cup \{\,*\,\}$ that uniquely associates each source feature with either a corresponding floating feature or the 'null' feature denoted $*$, thereby modelling the fact that some features may lack a correspondence in the floating set. In the usual case when Q is unknown, features are said to be unlabelled.

Let us further assume defined the *action* $T \star \mathbf{f}$ of any transformation $T \in \mathcal{T}$ on a source feature $\mathbf{f} \in I$, which depends on the geometrical nature of the chosen feature space. For features that represent 3D points, we simply have $T \star \mathbf{f} = T(\mathbf{f})$. Note, however, that the notation $T(\mathbf{f})$ is undefined for other types of features since T is a transformation from $\mathbb{R}^3$ as opposed to a transformation from $\mathcal{F}$.

In order to determine both the unknown spatial transformation T and the unknown correspondence function Q, we may adopt a variational approach whereby some measure of similarity between I and J is maximized with respect to both T and Q. It is natural to define the similarity measure as a sum of pairwise feature contributions:

$$S(T, Q) = \sum_k \text{sim}(T \star \mathbf{f}_k, Q(\mathbf{f}_k)), \tag{7.1}$$

where the sum involves all source features, including features with null correspondences, and $\text{sim} : \mathcal{F}^2 \to \mathbb{R}$ is a well-chosen function that quantifies the similarity of two associated features. We direct the reader to [10–12] for alternative variational formulations that make use of weighted multiple correspondences to reflect uncertainty about Q.

When features are directly comparable, it is intuitive to define the pairwise similarity function as a decreasing function of some distance in $\mathcal{F}$,

$$\text{sim}(\mathbf{f}, \mathbf{g}) = d_{\text{max}}^2 - d(\mathbf{f}, \mathbf{g})^2, \qquad \text{sim}(\mathbf{f}, *) = 0, \qquad (7.2)$$

where d_{max} is a cutoff parameter. In this simple instance, the pairwise similarity is upper bounded by d_{max}^2 and remains positive as long as the distance between associated features is less than d_{max}. More sophisticated similarity functions are possible and may be required for features that are not directly comparable, as we will see in Section 7.3 in the case of voxel features.

7.2.1 Generalized Iterative Closest Point Algorithm

Maximizing $S(T, Q)$ in Equation (7.1) is generally a non-convex problem that may require sophisticated numerical optimization procedures. In some cases, however, one may dramatically simplify the problem by performing an alternate maximization along T and Q, namely maximize $S(T, Q)$ with respect to Q at fixed T, then maximize $S(T, Q)$ with respect to T at fixed Q and then repeat until convergence. This alternate maximization scheme is guaranteed to find a local maximum of $S(T, Q)$ under mild regularity conditions, and this may be good enough in practice provided that a reasonable initial guess of T is available.

Specifically, alternate maximization leads to the following algorithm:

1. Initialize the transformation T and transform source features accordingly.
2. *Matching step* (maximization along Q). For each source feature $\mathbf{f}_k$, find the floating feature $\mathbf{g}_\ell$ most similar to $T \star \mathbf{f}_k$ according to $\text{sim}(T \star \mathbf{f}_k, \mathbf{g}_\ell)$. If $\text{sim}(T \star \mathbf{f}_k, \mathbf{g}_\ell) > 0$, set $Q(\mathbf{f}_k) = \mathbf{g}_\ell$; otherwise, discard the match, that is set $Q(\mathbf{f}_k) = *$.
3. *Registration step* (maximization along T). Update T so as to maximize the overall similarity between feature pairs $(T \star \mathbf{f}_k, Q(\mathbf{f}_k))$ tentatively labelled in Step 2 and transform source features accordingly.
4. Stop if both T and Q satisfy a convergence criterion. Return to Step 2 otherwise.

This scheme boils down to the well-known iterative closest point (ICP) algorithm [13–15] when sim(.) is the quadratic similarity defined in Equation (7.2) with $d(\mathbf{f}, \mathbf{g})$ being an Euclidean distance. The matching step (step 1) is a finite search that can be made efficient using K-D trees [14]. The registration step (step 2) is akin to a function-fitting problem and may have a closed-form solution depending on the chosen similarity function sim(.) and transformation search space $\mathcal{T}$. This is the case for the Euclidean similarity if $\mathcal{T}$ is the space of rigid transformations [16] or any linear space of transformations like affine or spline transformations [15]. When no closed-form solution exists, one has to resort to numerical optimization procedures such as gradient ascent [9] to perform Step 2.

7.2.2   Hierarchical Maximization

An alternative to the generalized ICP algorithm outlined in Section 7.2.1 is to tackle the maximization of Equation (7.1) using a hierarchical scheme, that is, consider the equivalent problem:

$$\max_{T \in \mathcal{T}} \bar{S}(T), \quad \text{where} \quad \bar{S}(T) \equiv \max_{Q} S(T, Q). \tag{7.3}$$

In this approach, the unknown correspondence function is 'optimized out' at fixed T; hence, the registration measure is defined as a function of T only. This involves performing the matching step (Step 2 of the generalized ICP outlined in Section 7.2.1) every time a new transformation T is tested, rather than relying on temporarily fixed correspondences as in Step 3 of the generalized ICP. This scheme may sometimes converge faster than the generalized ICP. Since the maximization problem stated in Equation (7.3) is most often non-convex, it however relies on the availability of robust numerical optimization procedures.

7.3   Intensity-Based Registration

Many registration methods proposed in the field of medical imaging are *intensity based* in that they use image intensity rather than geometrical features. A key motivation is to avoid performing a preliminary feature extraction, which may be challenging in practice. In intensity-based registration, features are voxels, and a pre-selection of voxels is generally not required. We now introduce intensity-based registration methods in historical context, starting with an intuitive example.

Suppose that it is desired to realign two images that are identical up to an unknown transformation. We understand that if we superimpose the images so as to perfectly align them, their intensities are identical everywhere in the region of overlap. Local discrepancies will be observed, however, if the images are misaligned (unless both images are uniform). Therefore, any positive-valued measure of intensity discrepancy, such as the sum of squared or absolute intensity differences, will cancel out when the images are in perfect match, hence reaching a global minimum.

Needless to say, the situation where images are perfectly identical up to a transformation does not occur in practice. But as long as they remain similar up to some 'noise', the simple principle described above proves very effective. Image difference-based registration methods can be traced back to the early 1970s [17, 18]. Difference measures, however, rely on the assumption that image intensities are directly comparable. When this condition is not met, for example due to different dynamic compression factors, it was soon suggested to use the correlation coefficient between image intensities as a registration measure to be maximized [19] since it is insensitive to affine intensity transformations.

Until the late 1980s, research focused on registration of images in the same modality. With the emergence of medical image processing, researchers started investigating multimodal registration problems such as magnetic resonance imaging/computed tomography (MRI/CT) or MRI/position emitted tomography (PET) for which early intensity difference or correlation techniques turned out ineffective. Work by Woods and colleagues [20] led to a significant advance in this direction. To overcome the assumption of affine dependence between image intensities, they proposed to partition one image into iso-intensity sets and build a registration measure based on intensity dispersion measures computed within each iso-intensity set projection in the other image. While the measure initially proposed by Woods was somewhat numerically unstable, it was later improved by several authors [21–23] and shown in [24] to closely relate with the correlation ratio, a classical non-linear extension of the correlation coefficient long used by statisticians.

Inspired by Woods *et al.*, Hill *et al.* introduced the useful concept of joint histogram, which counts intensity pairs induced by a transformation [25]. They observed that the joint histogram tends to 'disperse' with misregistration and thus proposed to measure registration in terms of joint histogram dispersion. Possible dispersion measures include the third-order moment, as originally proposed in [25], and entropy [26, 27], a well-known concept from information theory. It was then argued that such measures are not robust against partial image overlap, that is entropy tends to be minimized by transformations that disconnect the images. This observation motivated mutual information [28–31] and the entropy correlation coefficient, or normalized mutual information [31, 32], as registration measures to be maximized. An extensive comparison study of several multimodal rigid registration methods published in 1997 [33] established experimentally the effectiveness of mutual information registration.

In summary, research on multimodal image registration measures moved from the correlation coefficient to mutual information, two standard *statistical dependence* measures. An intuitive justification of both correlation-based and mutual information-based image registrations is that the degree of statistical dependence between intensities of associated voxels tends to be larger if voxels are matched than if they are not. This argument, however, lacks general theoretical justifications. Attempts were made by several authors to relate intensity-based registration measures to the standard statistical inference principle of maximum-likelihood estimation, however using rather restrictive image modelling assumptions [29, 34–39]. Such approaches are thus better interpreted in terms of *pseudo-likelihood* inference, as will be discussed thoroughly in Section 7.3.5.

7.3.1   Voxels as Features

We may formalize intensity-based registration as a special case of the feature-based registration framework outlined in Section 7.2, as proposed in [15]. In this view, features are voxels considered as four-dimensional points that concatenate spatial and intensity information. This means that each source feature $\mathbf{f}$ decomposes as a block vector:

$$\mathbf{f} = \begin{pmatrix} \mathbf{x} \\ i \end{pmatrix},$$

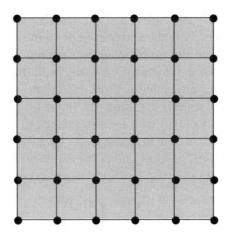

Figure 7.1   Dense spatial embedding of the floating image. Points within the image grid (in the shaded area) are interpolated, while points outside the image grid are assigned the unknown intensity symbol *

where $\mathbf{x}$ is a three-dimensional vector representing voxel coordinates in the given coordinate system associated with the source image I, and i is the corresponding voxel intensity. Likewise, each floating feature $\mathbf{g}$ decomposes as (Figure 7.1)

$$\mathbf{g} = \begin{pmatrix} \mathbf{y} \\ j \end{pmatrix},$$

where $\mathbf{y}$ and j, respectively, denote spatial coordinates in the coordinate system associated with the floating image J, and intensity.

To make things simple, we assume that the floating image J is a *dense* mapping from $\mathbb{R}^3$. This assumption can be realized by interpolating the input floating image using any three-dimensional image interpolation technique such as nearest neighbour, trilinear interpolation or higher order tensor B-spline interpolation [40–42]. However, interpolation is only defined within the floating grid. Guessing intensity outside the floating grid is an extrapolation problem that most likely cannot be solved accurately and thus makes little sense. Instead, we will assign an unknown intensity value $J(\mathbf{y}) =*$ to any point $\mathbf{y}$ that lies outside Ω_J, where $\Omega_J \subset \mathbb{R}^3$ is the convex hull of the floating grid.

To instantiate the general similarity measure defined in Equation (7.1), it is natural to define the pairwise feature similarity as a sum of spatial and intensity contributions:

$$\text{sim}(\mathbf{f}, \mathbf{g}) = -\lambda \|\mathbf{x} - \mathbf{y}\|^2 + s(i, j), \tag{7.4}$$

where $\|\mathbf{x} - \mathbf{y}\|$ denotes the Euclidean distance between voxel coordinates, λ is a positive-valued tradeoff parameter, and $s(i, j)$ is a pure intensity comparison function. In other words, two voxels are similar if both their locations and intensities are similar. While spatial similarity is measured by the negated Euclidean distance, intensity similarity is measured by some function $s(i,j)$ that should be chosen according to particular assumptions regarding the appearance of equivalent anatomical structures in both images, as will be discussed in greater details later in this chapter. We conventionally assume that the intensity similarity with the unknown intensity $*$ is zero, that is $s(i,*) = 0$ for any value i. From Equation (7.4), we observe that the larger the λ parameter, the smaller the similarity. In the limit that $\lambda \to \infty$, two voxels can have positive similarity only if their locations match exactly, that is $\mathbf{x} = \mathbf{y}$.

Assuming that intensities in I are invariant under spatial transformations, as is realistic if they represent physical *concentrations*, the action of any spatial transformation is intensity-preserving so that

$$T \star \begin{pmatrix} \mathbf{x} \\ i \end{pmatrix} = \begin{pmatrix} T(\mathbf{x}) \\ i \end{pmatrix},$$

for any source voxel with coordinates $\mathbf{x}$ and intensity i. The similarity measure then reads

$$S(T, Q) = \sum_k [-\lambda \|T(\mathbf{x}_k) - Q(\mathbf{x}_k)\|^2 + s(i_k, Q(i_k))], \tag{7.5}$$

where, using a slight notation abuse, $Q(\mathbf{x}_k)$ and $Q(i_k)$, respectively, denote the spatial coordinates and intensity of $Q(\mathbf{f}_k)$, that is

$$Q(\mathbf{f}_k) = \begin{pmatrix} Q(\mathbf{x}) \\ Q(i) \end{pmatrix},$$

We may tackle the problem of maximizing Equation (7.5) with respect to both the spatial transformation T and the correspondence function Q using either the generalized ICP algorithm outlined in Section 7.2.1, or the hierarchical maximization approach presented in Section 7.2.2. Both approaches require solving for the correspondence function Q by maximizing $S(T, Q)$ at fixed T. From Equation (7.5), this amounts to maximizing: $-\lambda \|T(\mathbf{x}_k) - Q(\mathbf{x}_k)\|^2 + s(i_k, Q(i_k))$ independently for each voxel k. This problem can be thought of as matching each transformed source voxel with the floating voxel most similar in intensity within a neighbourhood, the extent of which is inversely proportional to the λ parameter. As λ increases, the spatial constraint prevails and the best match gets closer to the trivial solution $Q(\mathbf{x}_k) = T(\mathbf{x}_k)$, meaning that, in the limiting case $\lambda \to \infty$, correspondences are fully determined by the transformation T regardless of image intensity information.

7.3.2 Special Case: Spatially Determined Correspondences

In the remainder of this chapter, we focus on the limiting case of spatially determined correspondences ($\lambda \to \infty$) as it is the most commonly used in practice. This is a situation where the generalized ICP algorithm is useless since it will always stick to the correspondences imposed by the initial transformation T, and will therefore be unable to make any move. In this case, we have to resort to the hierarchical maximization scheme presented in Section 7.2.2.

Since the optimal correspondences $Q_T^\star$ at fixed T verify $Q_T^\star(\mathbf{x}_k) = T(\mathbf{x}_k) + O(1/\lambda)$, the registration criterion for infinite λ simplifies to

$$\bar{S}(T) = \sum_k s(i_k, Q_T^\star(i_k)). \tag{7.6}$$

By definition, $Q_T^\star(i_k)$ is the (interpolated) floating image intensity at $Q_T^\star(\mathbf{x}_k)$, so we simply have $Q_T^\star(i_k) = J(T(\mathbf{x}_k))$ in this case. In the following, we abbreviate $J(T(\mathbf{x}_k))$ as $j_k^\downarrow$ to simplify mathematical notation. Although this notation is somewhat abusive, it should be clear from the context that $j_k^\downarrow$ is a function of T. We also denote by $\mathcal{A}^\downarrow$ the set of source voxel indices k such that $T(\mathbf{x}_k)$ falls at an interpolable location. Since $s(i_k, *) = 0$ for any $k \notin \mathcal{A}^\downarrow$, the above Equation (7.6) thus reads

$$\bar{S}(T) = \sum_{k \in \mathcal{A}^\downarrow} s(i_k, j_k^\downarrow), \tag{7.7}$$

and defines a general registration function to be maximized over the transformation search space. Specific instantiations of Equation (7.7) will follow from particular choices for the intensity similarity function $s(i, j)$.

7.3.3 Intensity Difference Measures

When equivalent structures are assumed to have similar intensities in both the source and floating images, a natural choice for the similarity function $s(i, j)$ is a negated distance in $\mathbb{R}$:

$$s(i, j) = -|i - j|^\alpha,$$

where α is a strictly positive value. The registration function $\bar{S}(T)$ then reads

$$\bar{S}(T) = - \sum_{\mathcal{A}^\downarrow} |i_k - j_k^\downarrow|^\alpha,$$

therefore maximizing $\bar{S}(T)$ amounts to minimizing the sum of absolute intensity differences raised to the power α. Common choices are $\alpha = 2$, leading to the sum of squared differences (SSD) measure, and $\alpha = 1$, leading to the sum of absolute Differences (SAD) measure. The smaller α, the more 'robust' the registration measure in the sense that it is less sensitive to large intensity differences. Choosing small α values may be appropriate when occlusions are present, for instance due to evolving tumours or lesions.

A more general approach to define robust registration functions is to modulate intensity differences using a slowly increasing ρ-function by analogy with M-estimators [23]:

$$\bar{S}(T) = - \sum_{\mathcal{A}^\downarrow} \rho(i_k - j_k^\downarrow)$$

Possible choices for the ρ-function include

- The Huber ρ-function:

$$\rho(x) = \min\ (x^2,\ 2c|x| - c^2)$$

- The Cauchy ρ-function:

$$\rho(x) = c^2 \log\ [1 + (\tfrac{x}{c})^2]$$

- The Geman–McClure ρ-function:

$$\rho(x) = \frac{x^2}{1 + (x/c)^2}$$

- The Tukey ρ-function:

$$\rho(x) = \frac{c^2}{3}\min\ (1 - [1 - (x/c)^2]^3,\ 1)$$

These examples, like many other ρ-functions, involve a parameter c to be tuned, which plays the role of a cutoff distance in the sense that intensity differences larger than c are downweighted compared to the SSD measure, which corresponds to a quadratic ρ-function.

7.3.4   Correlation Coefficient

Intensity difference registration measures have two important drawbacks. First, they rely on an implicit assumption of intensity conservation that may not hold, even in monomodal registration problems, if, for instance, the source and floating images have different intensity dynamic ranges. Second, they are highly overlap-dependent and are therefore prone to strong spurious local maxima. This can easily be seen by considering the registration of an image onto itself; since any transformation for which intensity differences are all zero in the image overlap is optimal in the sense of the SSD or any other pure intensity difference measure, the optimization may pick an erroneous transformation, such as one that maps background to background.

A similarity function that has the potential to better deal with both lack of intensity conservation and overlap dependence is the product of normalized intensities:

$$s(i,j) = \frac{(i - \mu_I)(j - \mu_J)}{\sigma_I \sigma_J}, \tag{7.8}$$

where μ_I, σ_I, μ_J and σ_J are mean and standard deviation parameters, respectively, for the source and floating image. In practice, they can be estimated before-hand as the empirical intensity means and standard deviations computed from the whole images. However, it is common practice to restrict their estimation to the image overlap at a given transformation T, thereby computing on-the-fly estimates:

$$\hat{\mu}_{IT} = \frac{1}{n_T} \sum_{A^\downarrow} i_k, \qquad \hat{\sigma}_{IT} = \sqrt{\frac{1}{n_T} \sum_{A^\downarrow} (i_k - \hat{\mu}_{IT})^2},$$

and, likewise,

$$\hat{\mu}_{JT} = \frac{1}{n_T} \sum_{A^\downarrow} j_k^\downarrow, \qquad \hat{\sigma}_{JT} = \sqrt{\frac{1}{n_T} \sum_{A^\downarrow} (j_k^\downarrow - \hat{\mu}_{JT})^2},$$

where $n_T = \#A^\downarrow$ denotes the overlap size, that is the number of source voxels with an interpolable floating correspondent by T.

Plugging these estimates into Equation (7.8), the similarity function itself becomes a function of T:

$$s_T(i,j) = \frac{(i - \hat{\mu}_{IT})(j - \hat{\mu}_{JT})}{\hat{\sigma}_{IT}\hat{\sigma}_{JT}},$$

and we may approximate the registration measure by

$$\tilde{S}(T) = \sum_{A^\downarrow} \frac{(i_k - \hat{\mu}_{IT})(j_k^\downarrow - \hat{\mu}_{JT})}{\hat{\sigma}_{IT}\hat{\sigma}_{JT}}, \qquad (7.9)$$

where the tilde used in the notation is to stress the distinction between the plug-in registration measure $\tilde{S}(T)$ and the registration measure $\bar{S}(T)$ that could be computed if μ_I, σ_I, μ_J and σ_J were known in advance.

Interestingly, $\tilde{S}(T)$ as defined in Equation (7.9) also reads $\tilde{S}(T) = n_T \hat{\rho}_T$, where $\hat{\rho}_T$ is the Pearson correlation coefficient of source and floating intensities computed in the image overlap. This implies that maximizing $\tilde{S}(T)$ is roughly equivalent to maximizing the correlation coefficient subject to a 'large overlap' constraint imposed by the overlap size term, n_T. Most existing implementations of correlation-based registration seem to drop n_T in practice; however, it is useful to downweight transformations corresponding to small overlap and incidentally large correlation.

7.3.5 Pseudo-likelihood Measures

The correlation measure discussed in Section 7.3.4 is most often appropriate for monomodal image registration problems, where the two images I and J to be registered can be assumed to be approximately related by an affine intensity mapping, that is loosely speaking, there exist constants α and β such that $I \approx \alpha \, J{\circ}T + \beta$ for the correct transformation. While affine intensity dependence is a weaker assumption than intensity conservation, which corresponds to the special case $\alpha = 1$ and $\beta = 0$, it is unlikely to hold for images acquired from different modalities (if it does, then the complementarity of image modalities is questionable). Therefore, correlation-based registration generally yields poor results in multimodal image registration.

To define registration measures that are appropriate for multimodal registration, it is necessary to rely on more general intensity dependence models. This goal can be approached using a statistical inference procedure, which requires defining a probabilistic image dependence model parametrized by the unknown transformation, and an associated estimation strategy for the transformation.

Specifically, let us model the matched intensity pairs $(i_k, j_k^\downarrow)$ as being independently and identically distributed according to a known probability density function $p(i, j)$. This provides us with a simplistic joint image formation model, which does not account for spatial dependencies and non-stationarities in the images, and should therefore not be considered to be the *true* sampling model in a frequentist sense. We may instead consider it as a *working model* or an *encoder* in the terminology of minimum description length model selection [43]. According to the working model, the joint probability of matched intensity pairs is given by

$$P((i_1, j_1^\downarrow), (i_2, j_2^\downarrow), \ \ldots \ , (i_{n_T}, j_{n_T}^\downarrow) | T) = \prod_k p(i_k, j_k^\downarrow), \tag{7.10}$$

under the assumption that T is a one-to-one spatial mapping, meaning that T maps distinct points to distinct points. Equation (7.10), however, does not define a valid likelihood function because it describes the probability distribution of a *subset* of the data that depends on T, rather than the probability distribution of the full data (I, J). In order to specify a distribution $P(I, J | T)$ that is compatible with Equation (7.10), we may apply the PDF projection theorem [44], which implies that the following quantity:

$$P(I, J | T) = \Pi(I)\Pi(J) \prod_{A^\downarrow} \frac{p(i_k, j_k^\downarrow)}{\pi(i_k)\pi(j_k^\downarrow)}, \tag{7.11}$$

defines a valid joint probability distribution for the input images. In Equation (7.11), $\Pi(I) = \prod_k \pi(i_k)$ and $\Pi(J) = \prod_\ell \pi(j_\ell)$ are *arbitrary* marginal distributions for I and J, respectively, which we also assume to factorize across voxels for computational tractability. The PDF projection theorem is, in fact, an application of the minimum Kullback–Leibler divergence principle or maximum relative entropy principle; given a reference distribution $\Pi(I)\Pi(J)$ that models the images as spatially unrelated, it selects the distribution $P(I, J | T)$ that is closest to $\Pi(I)\Pi(J)$ according to the Kullback–Leibler divergence, subject to the constraint provided by Equation (7.10), which encodes some spatial dependence between the images.

Equation (7.11) defines a *pseudo-likelihood* function in that it specifies the data distribution according to a simplistic rather than physically plausible image formation model. Since the marginals $\Pi(I)$ and $\Pi(J)$ are independent from T, they play no role in registration and can be dropped. Taking the logarithm of Equation (7.11), we see that maximizing the pseudo-likelihood with respect to T is equivalent to maximizing the general registration measure in Equation (7.7) with the specific similarity function:

$$s(i, j) = \log \frac{p(i, j)}{\pi(i)\pi(j)},$$

leading to the pseudo-log-likelihood registration measure

$$\bar{S}(T) = \sum_{A^{\downarrow}} \log \frac{p(i_k, j_k^{\downarrow})}{\pi(i_k)\pi(j_k^{\downarrow})}, \tag{7.12}$$

For practical implementation, we need to estimate the joint intensity distribution $p(i,j)$ as well as to pick sensible reference distributions $\pi(i)$ and $\pi(j)$. In a *supervised learning* approach, we may opt to estimate $p(i,j)$ from a training set of similar image pairs [45]. It is then natural to choose the reference distributions as the respective marginals corresponding to $p(i,j)$,

$$\pi(i) = \int p(i,j)dj, \qquad \pi(j) = \int p(i,j)di. \tag{7.13}$$

However, training data may not be available in practice and we may then have to resort to an *unsupervised* approach whereby $p(i,j)$, $\pi(i)$ and $\pi(j)$ are estimated on-the-fly from the image pair at hand. We discuss in the following different distribution models and associated estimation strategies.

7.3.5.1 Correlation Revisited

As a simple approach, let us assume that $p(i,j)$ is a bivariate Gaussian distribution with unknown mean vector $(\mu_I \ \mu_J)^{\mathsf{T}}$ and unknown 2-by-2 covariance matrix Σ,

$$p(i,j) = \frac{1}{2\pi|\Sigma|} \exp\left(-\frac{1}{2}(i - \mu_I \ j - \mu_J)\Sigma^{-1}(i - \mu_I \ j - \mu_J)^{\mathsf{T}}\right).$$

In order to eliminate the unknown parameters, we may take the maximum of the pseudo-log-likelihood in Equation (7.12) with respect to μ_I, μ_J and Σ. This is in essence the idea behind the classical concept of *profile likelihood* in statistical estimation. The optimal parameters according to this criterion turn out to be the familiar empirical mean and covariance matrix computed in the image overlap, yielding

$$\hat{\mu}_{IT} = \frac{1}{n_T} \sum_{A^{\downarrow}} i_k, \qquad \hat{\mu}_{JT} = \frac{1}{n_T} \sum_{A^{\downarrow}} j_k^{\downarrow},$$

for the mean and

$$\hat{\Sigma}_T = \frac{1}{n_T} \sum_{A^{\downarrow}} (i_k - \hat{\mu}_{IT} \ j_k^{\downarrow} - \hat{\mu}_{JT})^{\mathsf{T}}(i_k - \hat{\mu}_{IT} \ j_k^{\downarrow} - \hat{\mu}_{JT}),$$

for the covariance matrix. Note that the estimated covariance matrix may be rewritten as

$$\hat{\Sigma}_T = \begin{pmatrix} \hat{\sigma}_{IT}^2 & \hat{\rho}_T\hat{\sigma}_{IT}\hat{\sigma}_{JT} \\ \hat{\rho}_T\hat{\sigma}_{IT}\hat{\sigma}_{JT} & \hat{\sigma}_{JT}^2 \end{pmatrix},$$

where $\hat{\sigma}_{IT}$ and $\hat{\sigma}_{JT}$ are the empirical standard deviations and $\hat{\rho}_T$ is the empirical Pearson correlation coefficient already considered in Section 7.3.4.

To instantiate the reference distributions $\pi(i)$ and $\pi(j)$ in Equation (7.12), we may take the marginals of $p(i,j)$ as in Equation (7.13). These are seen to be univariate Gaussian distributions with parameters $(\hat{\mu}_{IT}, \hat{\sigma}_{IT}^2)$ and $(\hat{\mu}_{JT}, \hat{\sigma}_{JT}^2)$, respectively. Note that this is equivalent to *minimizing* the pseudo-likelihood measure with respect to the parameters of univariate Gaussian distribution models for $\pi(i)$ and $\pi(j)$. In other words, our distribution estimation strategy approximates the pseudo-likelihood as a ratio of maximum likelihoods, similar to the *maximum-likelihood ratio* test used in statistical hypothesis testing.

Plugging the distribution estimates into the registration measure, we obtain the expression of the profile log-likelihood after some basic mathematical manipulation:

$$\tilde{S}(T) = -\frac{n_T}{2} \log (1 - \hat{\rho}_T^2). \tag{7.14}$$

This measure is an increasing function of the squared correlation coefficient and is positive-valued because $\hat{\rho}_T$ varies in the range $[-1, 1]$. It is comparable to the overlap-normalized correlation measure, $n_T\hat{\rho}_T$, derived in Section 7.3.4. However, while both the pseudo-likelihood measure in Equation (7.14) and $n_T\hat{\rho}_T$ essentially encourage large correlation and large image overlap, their respective maximizations are not equivalent for two reasons. First, the pseudo-likelihood measure is insensitive to the sign of $\hat{\rho}_T$, and may thus rate well transformations that lead to large negative intensity correlation. Second, because the pseudo-likelihood measure is superlinear in $\hat{\rho}_T$, it is also less dependent on the overlap size n_T than $n_T\hat{\rho}_T$.

7.3.5.2 Mutual Information

The Gaussian joint intensity distribution model implicitly assumes an affine dependence between image intensities, and leads to a variant of correlation that, as discussed earlier, is only appropriate for monomodal image registration in practice. However, the pseudo-likelihood framework enables us to derive registration measures that can deal with more complex types of intensity dependence.

We now turn to a piecewise constant model for $p(i,j)$ constructed as follows. Let us consider two partitions of $\mathbb{R}$ in intervals, denoted $U_{a=1,2,\ldots,\infty}$ and $V_{b=1,2,\ldots,\infty}$. For clarity, we assume intervals of constant sizes α and β, respectively, but this is not essential. Let us then model $p(i,j)$ as

$$p(i,j) = \frac{1}{\alpha\beta} \sum_{a,b} P_{ab} \chi_{ab}(i,j),$$

where χ_{ab} is the indicator function of $U_a \times V_b$, and P_{ab} are unknown parameters. By construction, P_{ab} is the probability that (i,j) falls within the bin $U_a \times V_b$, therefore, we have

$$\forall(a, b), \quad P_{ab} \geq 0, \quad \sum_{a,b} P_{ab} = 1.$$

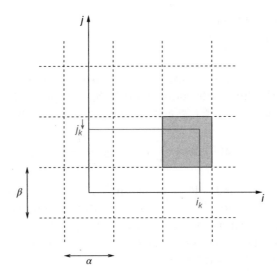

Figure 7.2   Joint intensity binning

Similarly, we use a piecewise constant model for the marginal reference distributions:

$$\pi(i) = \frac{1}{\alpha} \sum_a \Pi_a^i \chi_a(i), \qquad \pi(j) = \frac{1}{\beta} \sum_b \Pi_b^j \chi_b(j),$$

where Π_a^i and Π_b^j hence represent the binned probabilities associated with $\pi(i)$ and $\pi(j)$, respectively.

Under this piecewise constant joint intensity distribution model, we can rewrite the pseudo-log-likelihood measure from Equation (7.12) as (Figure 7.2)

$$\bar{S}(T) = \sum_{a,b} h_{abT} \log \frac{P_{ab}}{\Pi_a^i \Pi_b^j},$$

where h_{abT} is the empirical count of intensity pairs $(i_k, j_k^{\downarrow})$ within $U_a \times V_b$. In the image registration literature, h_{abT} is referred to as the *joint image histogram*. It should be kept in mind that this is a function of T, and that its definition depends on an underlying partition of the intensity space.

As in Section 7.3.5.1, we may eliminate the unknown parameters by maximizing the pseudo-log-likelihood at fixed T with respect to P_{ab} and minimizing it with respect to Π_a^i and Π_b^j. Let

$$H_{abT} = \frac{1}{n_T} h_{abT}$$

denote the normalized joint histogram, and H_{aT} and H_{bT} denote the marginals associated with H_{abT}:

$$H_{aT} = \sum_b H_{abT}, \qquad H_{bT} = \sum_a H_{abT}.$$

The solution is found to be

$$\hat{P}_{abT} = H_{abT}, \qquad \hat{\Pi}^i_{aT} = H_{aT}, \qquad \hat{\Pi}^j_{bT} = H_{bT}.$$

While there is a conceptual distinction to make between H_{abT}, an empirical distribution, and P_{ab}, the distribution model, H_{abT} happens to be the maximum pseudo-likelihood estimate of P_{ab} and the plug-in pseudo-log-likelihood estimate reads

$$\tilde{S}(T) = n_T \sum_{a,b} H_{abT} \log \frac{H_{abT}}{H_{aT} H_{bT}}. \tag{7.15}$$

The sum in the right-hand side of Equation (7.15) is recognized to be the *mutual information*, hereafter denoted $\mathcal{I}_T$, associated with the empirical distribution H_{abT}. $\mathcal{I}_T$ is a standard information-theoretic measure [46] of the amount of information that source and floating binned image intensities convey about one another. Equation (7.15) may be compactly rewritten as

$$\tilde{S}(T) = n_T \mathcal{I}_T.$$

Hence, maximizing the profile pseudo-likelihood with respect to T amounts to maximizing mutual information weighted by the overlap size. As in the case of correlation-based registration, few implementations of mutual information registration seem to consider n_T in practice. It was also suggested [32] to replace mutual information with a variant called *normalized mutual information*, which proves empirically to be less sensitive to changes in image overlap. The normalized mutual information measure is defined as

$$\left(1 - \frac{\mathcal{I}_T}{\hat{\mathcal{H}}_{IT} + \hat{\mathcal{H}}_{JT}} \right)^{-1},$$

where $\hat{\mathcal{H}}_{IT}$ and $\hat{\mathcal{H}}_{JT}$ are the marginal entropies associated with H_{abT},

$$\hat{\mathcal{H}}_{IT} = - \sum_a H_{aT} \log H_{aT}, \qquad \hat{\mathcal{H}}_{JT} = - \sum_b H_{bT} \log H_{bT}.$$

Figure 7.3 compares the different overlap normalization strategies for mutual information, showing the effect of applying large translations to a source image, thereby inducing massive changes in overlap size. Also note the rugged aspect of the measure profiles, which is an effect of interpolating the floating image and is difficult to avoid [47–51].

Mutual information formalizes the intuitive idea that the statistical dependence between associated voxels tends to increase as T approaches the correct registration transformation. The joint histogram shows more 'compactness' or more 'coherence' as mutual information increases, and it is assumed that the more coherence, the better registration. This is illustrated in Figures 7.4 and Figures 7.5 in monomodal and multimodal registration cases, respectively.

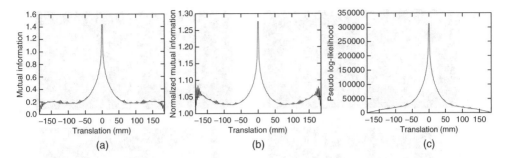

Figure 7.3 Profiles of three variants of the mutual information measure against translations with constant direction. (a) Mutual information, (b) normalized mutual information and (c) pseudo-log-likelihood

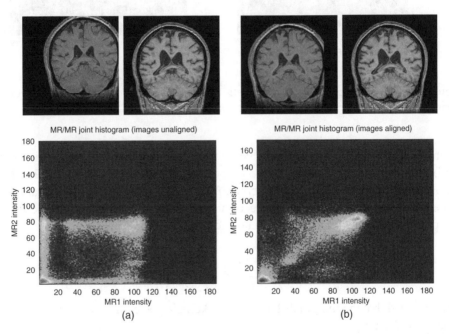

Figure 7.4 Joint histogram of two brain magnetic resonance volumes using 256 bins for both images. (a) Unregistered images and (b) registered images

7.3.5.3 Kernel Pseudo-likelihood

As a natural extension of the piecewise constant model, we may estimate the joint intensity distribution at fixed T using the kernel method, also known as the Parzen window method:

$$\hat{p}_T(i,j) = \frac{1}{n_T} \sum_{\mathcal{A}^{\downarrow}} \psi(i - i_k, \, j - j_k^{\downarrow}), \tag{7.16}$$

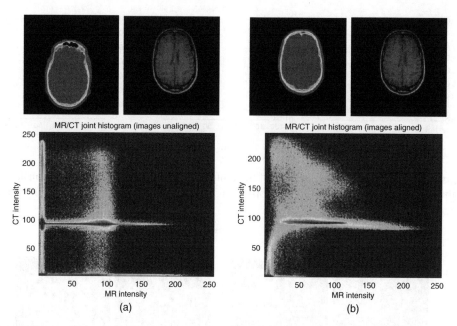

Figure 7.5 Joint histogram of brain magnetic resonance and computed tomography volumes using 256 bins for both images. (a) unregistered images and (b) registered images

where $\psi : \mathbb{R}^2 \to \mathbb{R}$ is a positive-valued kernel that integrates to one. This defines a mixture distribution with as many components as observed intensity pairs. Upon choosing a smooth (e.g. Gaussian) kernel, $\hat{p}_T$ is a smooth distribution contrary to the piecewise constant model used in Section 7.3.5.2. Equation (7.16) may be used to compute a plug-in estimate of the pseudo-log-likelihood:

$$\tilde{S}(T) = \sum_{A^\downarrow} \log \frac{\hat{p}_T(i_k, j_k^\downarrow)}{\hat{p}_T(i_k)\hat{p}_T(j_k^\downarrow)}, \tag{7.17}$$

being noticed that the marginals have similar kernel-type expressions as the joint distribution estimate:

$$\hat{p}_T(i) = \frac{1}{n_T} \sum_{A^\downarrow} \psi_i(i - i_k), \qquad \hat{p}_T(j) = \frac{1}{n_T} \sum_{A^\downarrow} \psi_j(j - j_k^\downarrow),$$

where $\psi_i(i) = \int \psi(i, j)dj$ and $\psi_j(j) = \int \psi(i, j)di$.

We note that $\hat{p}_T$ generally does not maximize the pseudo-likelihood on the space of mixture distributions, so that Equation (7.17) cannot be interpreted as a maximum-likelihood ratio. Nevertheless, the kernel pseudo-likelihood measure mitigates the binning effect of using a piecewise constant distribution model, which may be responsible for spurious local maxima in the mutual information registration measure, as illustrated with a synthetic experiment in Figures 7.6 and 7.7.

Figure 7.6 'Grey stripe' images. The floating is a binary image with size 40 × 30 pixels. The source is a gradation image with size 30 × 30 pixels in which each column has a different intensity. (a) Floating image and (b) source image

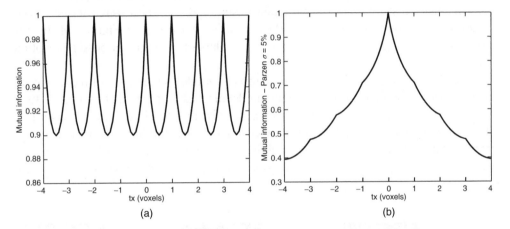

Figure 7.7 Profiles of mutual information and kernel pseudo log-likelihood against horizontal translations in the 'grey stripe' registration experiment. Mutual information is globally maximal for any integer translation (for non-integer translations, smaller values are observed due to floating intensity interpolation). (a) Mutual information and (b) Kernel pseudo-likelihood

7.3.5.4 Finite Mixture Pseudo-likelihood

While Section 7.3.5.3 discussed a virtually infinite mixture model for the joint intensity distribution based on the kernel method, we may also consider a finite mixture model [35, 52],

$$p(i,j) = \sum_{c=1}^{C} w_c N((i\ j)^{\mathsf{T}}; \mu_c, \Sigma_c),$$

where C is a fixed number of components, $w_c \in [0, 1]$ is a proportion and $N(.; \mu_c, \Sigma_c)$ denotes the normal bivariate distribution with mean μ_c and covariance matrix Σ_c. In order to estimate the mixture model parameters w_c, μ_c and Σ_c, for $c = 1, \ldots, C$, we may again maximize the pseudo-likelihood measure in Equation (7.12), generally leading to a non-convex problem with no explicit solution. A local maximizer may, however, be computed using a numerical iterative scheme such as the expectation-maximization (EM) algorithm [53].

Regarding the reference distributions $\pi(i)$ and $\pi(j)$, it is natural to model them as univariate Gaussian mixture models with C classes, and fit the parameters using similar numerical procedures as for the joint distribution. Another, simpler, approach is to use single Gaussian models, in which case the maximum-likelihood parameter estimates are given by the empirical means and variances $\hat{\mu}_{IT}$, $\hat{\sigma}^2_{IT}$, $\hat{\mu}_{JT}$, $\hat{\sigma}^2_{JT}$, as in Section 7.3.5.1.

Interestingly, the EM algorithm provides us not only with mixture parameter estimates but also with posterior probabilities, for each voxel in the image overlap, of belonging to one of the C mixture components. Interpreting each component as the joint intensity distribution of a particular tissue type, we get a rough tissue classification as a by-product of image registration, as illustrated in Figure 7.8. In this sense, finite mixture pseudo-likelihood is a collaborative registration-segmentation approach.

An interesting special case of the finite mixture model is when one of the images to be registered, say image J, is a sufficiently detailed label image that assigns distinct labels to the tissues that have distinct intensity distributions in the other image. It is then appropriate to choose C as the number of labels in J and force the variance σ^2_j to zero. In this special case, the component index c can be confused with the intensity j in the floating image, and the maximum-likelihood parameter estimates have a simple

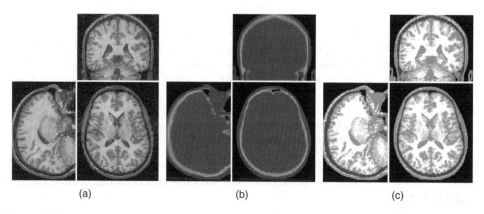

(a)   (b)   (c)

Figure 7.8 Example of rigid CT-to-MRI registration using finite mixture pseudo-likelihood with $C=5$ components. Shown in panel (c) is the final maximum a posteriori tissue classification using the outcome of the EM algorithm. The CT information makes it possible to segment the skull, which is indistinguishable in intensity from brain tissues in the MRI volume. (a) MRI volume, (b) registered CT volume, and (c) 5 class segmentation

closed-form expression,

$$\hat{\mu}_j = \frac{1}{n_{jT}} \sum_{A_j^\downarrow} i_k, \qquad \hat{\Sigma}_j = \begin{pmatrix} \frac{1}{n_{jT}} \sum_{A_j^\downarrow} (i_k - \hat{\mu}_j)^2 & 0 \\ 0 & 0 \end{pmatrix}, \qquad \hat{\pi}_j = \frac{n_{jT}}{n_T},$$

where $A_j^\downarrow$ is the subset of source voxel indices k that transform to the floating iso-intensity set with level j, that is $A_j^\downarrow = \{k \in A^\downarrow, \ j_k^\downarrow = j\}$, and $n_{jT} = \#A_j^\downarrow$ its cardinal number.

Plugging these estimates in Equation (7.12), and using a single Gaussian model for $\pi(i)$, we obtain the following expression for the plug-in pseudo-log-likelihood function,

$$\tilde{S}(T) = -\frac{n_T}{2} \log (1 - \hat{\eta}_T^2), \tag{7.18}$$

where $\hat{\eta}_T$ is the empirical *correlation ratio* of intensities in I and J, defined as

$$\hat{\eta}_T^2 = 1 - \frac{1}{n_T \hat{\sigma}_J^2} \sum_j \sum_{A_j^\downarrow} (i_k - \hat{\mu}_j)^2$$

The correlation ratio is a standard statistical coefficient that measures the degree of *non-linear* functional dependence between two variables. It is an asymmetrical measure in that the two variables do not play the same role; in this case, the source image I is the predicted variable and the floating image J is the predictor; however, the above reasoning remains valid when reversing the roles. Similar to the correlation coefficient, which measures affine dependence, the correlation ratio takes on values between 0 and 1, and is always larger than the correlation coefficient:

$$0 \le \hat{\rho}_T^2 \le \hat{\eta}_T^2 \le 1.$$

Note the striking similarity between Equation (7.18) and the normalized correlation coefficient measure in Equation (7.14), which corresponds to the plug-in pseudo-log-likelihood measure associated with a Gaussian joint intensity distribution model. Both measures relate registration quality with textbook statistical dependence indices long used by statisticians.

7.3.6 General Implementation Using Joint Histograms

When deriving the mutual information measure in Section 7.3.5.2, we introduced the concept of joint histogram, which involves binning intensities in both the source and floating domains, and simply counting the number of intensity pairs in each bin:

$$h_{abT} = \#\{k \in A^\downarrow, \ \chi_{ab}(i_k, j_k^\downarrow) = 1\},$$

where $a = 1, 2, \ldots$ and $b = 1, 2, \ldots$ are bin indices.

Table 7.1 Some popular medical image registration measures expressed in terms of the normalized joint histogram H_{abT} (note that the dependence in T is omitted from the formulas for clarity). Measures marked with a star (*) are to be minimized. Measures marked with a dag ($\dagger$) are not invariant through swapping the rows and columns of the joint histogram

Dependence type	Measure	Formula
Intensity conservation	Mean square difference*	$\sum_{a,b} H_{ab}(a-b)^2$
	Mean absolute difference*	$\sum_{a,b} H_{ab}\lvert a-b\rvert$
Linear	Normalized inter-correlation	$\sum_{a,b} H_{ab}\dfrac{ab}{e_A e_B}$
	Alpert criterion* $\dagger$	$\sum_{a,b} H_{ab}\left(\dfrac{\mu_B}{\mu_A}a-b\right)^2$
Affine	Correlation coefficient	$\sum_{a,b} H_{ab}\dfrac{(a-\mu_A)(b-\mu_B)}{\sigma_A\sigma_B}$
Functional	Woods criterion* $\dagger$	$\sum_b H_b\dfrac{\sigma_b}{\mu_b}$
	Correlation ratio$\dagger$	$1-\dfrac{1}{\sigma_A^2}\sum_b H_b\sigma_b^2$
General	Mutual information	$\sum_{a,b} H_{ab}\log\dfrac{H_{ab}}{H_a H_b}$
	Kernel mutual information	$\sum_{a,b} H_{ab}\log\dfrac{\tilde{H}_{ab}}{\tilde{H}_a \tilde{H}_b}$
	Normalized mutual information	$\dfrac{\sum_{ab} H_{ab}\log H_{ab}}{\sum_a H_a\log H_a+\sum_b H_b\log H_b}$

Notations $\quad H_a=\sum_b H_{a,b}, \quad H_b=\sum_a H_{a,b}, \quad H_{a\lvert b}=\frac{H_{ab}}{H_b}\quad (=0\text{ if }H_b=0)$

$\mu_A=\sum_a H_a\,a, \quad e_A^2=\sum_a H_a\,a^2, \quad \sigma_A^2=e_A^2-\mu_A^2$

$\mu_B=\sum_b H_b\,b, \quad e_B^2=\sum_b H_b\,b^2, \quad \sigma_B^2=e_B^2-\mu_B^2$

$\mu_b=\sum_a H_{a\lvert b}\,a, \quad \sigma_b^2=\sum_a H_{a\lvert b}\,a^2-\mu_b^2$

$\tilde{H}_{ab}=G_{ab}\star H_{ab}, \quad \tilde{H}_a=G_a\star H_a, \quad \tilde{H}_b=G_b\star H_b$

(2D and 1D convolutions by filters G_{ab}, G_a and G_b)

The joint histogram is thus a two-dimensional array which encodes, up to the binning, the whole empirical distribution of intensity pairs induced by a particular spatial transformation. It turns out to be a useful computational step to implement not only the mutual information measure but also many other intensity-based registration measures, which makes it attractive for a general intensity-based registration implementation framework. Table 7.1 lists some of the most widely used intensity-based registration measures in medical imaging, and provides their mathematical expression depending on the normalized joint histogram, defined as

$$H_{abT}=\frac{h_{abT}}{n_T}=\frac{h_{abT}}{\sum_{a,b} h_{abT}}.$$

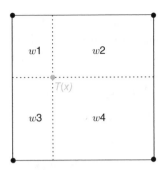

Figure 7.9 2D representation of trilinear interpolation weights. The center dot represents a transformed voxel. Outer dots represent grid neighbours

Although intensity binning entails approximations, approximation errors can be made negligible by choosing sufficiently small bin sizes. It is a common practice to use fixed number of bins that correspond to multiples of two, for example 64×64, 128×128, 256×256, ... Note that there also exist adaptive approaches [54] that make bin sizes depend on the image overlap size.

A difficulty that arises in the context of optimizing a registration measure is that the joint histogram, as defined earlier, is not a continuous function of T due to binning. As T varies, the intensity $j_k^\downarrow$ corresponding to a given source voxel may change bin, hence causing abrupt changes in bin counts. As a consequence, the joint histogram is a discontinuous function of the transformation parameters, and so will be any registration measure computed from the joint histogram.

This, in practice, precludes the use of standard numerical optimization methods that typically assume a smooth objective function. One way to work around this problem is to replace floating intensity interpolation with another interpolation scheme that guarantees smooth variations of the joint histogram. The partial volume (PV) interpolation method [28, 31] is one example of such a scheme that expands on the trilinear interpolation method as follows.

In trilinear interpolation, we interpolate the intensity of any transformed point $T(\mathbf{x}_k)$ as a weighted sum of its eight floating grid neighbours (see Figure 7.9):

$$j_k^\downarrow = \sum_\ell w_{k\ell}^\downarrow j_\ell,$$

where $w_{k\ell}^\downarrow$ are transformation-dependent positive weights that sum up to one. Specifically, we have for each grid neighbour $\mathbf{y}_\ell$

$$w_{k\ell}^\downarrow = \beta_3^1(\Delta^{-1}(T(\mathbf{x}_k) - \mathbf{y}_\ell)),$$

where Δ is the diagonal matrix with the floating voxel sizes in each direction as its diagonal elements, and β_3^1 is the order 1 tensor-product B-spline function, that is the coordinate-wise product of triangle functions:

$$\beta_3^1(\mathbf{x}) = \prod_{n=1}^{3} \max(0, 1 - |x_n|).$$

The idea of PV interpolation is to avoid interpolating $j_k^\downarrow$ and, instead, distribute over the joint histogram the eight intensity pairs (i_k, j_ℓ) that correspond to the floating grid neighbours of $T(\mathbf{x}_k)$. This is done by incrementing each bin where (i_k, j_ℓ) falls proportionally to the trilinear interpolation weight $w_{k\ell}^\downarrow$, implying that the joint histogram is actually computed using the following double sum:

$$h_{abT} = \sum_{k \in A^\downarrow} \sum_{\ell} w_{k\ell}^\downarrow \chi_{ab}(i - i_k, j - j_\ell). \tag{7.19}$$

Note that the PV interpolation method also makes it possible to partially account for points that transform within one voxel outside the floating grid domain, by simply ignoring their neighbours with non-existing floating intensities. This way, the overlap size n_T computed as the total joint histogram mass is a non-integer real value that varies smoothly with the transformation parameters.

Several variants of PV interpolation have been proposed in the literature such as modulating the trilinear interpolation weights by the intensity [55] or randomly selecting a single neighbour with probability equal to its weight [48]. We refer the reader to [47–51] for several experimental studies of the impact of joint histogram interpolation on image registration, and more proposals regarding interpolation schemes.

A final remark regarding the implementation of joint histogram-based registration measures is that it is possible, and often desirable in practice, to sub-sample the source image in order to speed up computation. Since the complexity of computing the joint histogram is proportional to the number of voxels in the source image, applying for instance a regular sub-sampling with factors $2 \times 2 \times 2$ will result in cutting down computation time by a factor 8. Other options than regular sub-sampling include random sampling [56] or preferential selection of voxels near image contours [57].

7.4   Transformation Spaces and Optimization

We now turn to the problem of optimizing a registration measure $\tilde{S}(T)$ such as those described in Section 7.3 over a space $\mathcal{T}$ that describes the set of physically plausible spatial transformations in a particular registration problem. Generally speaking, the optimization procedure involves representing members of $\mathcal{T}$ by a vector-valued free parameter $\theta \in \mathbb{R}^d$ that defines an onto mapping $\theta \mapsto T_\theta$ from $\mathbb{R}^d \to \mathcal{T}$. The problem then amounts to maximizing the registration measure $\tilde{S}(\theta) = \tilde{S}(T_\theta)$ over $\mathbb{R}^d$.

To that end, standard numerical optimization procedures can be used, such as quasi-search methods [54, 58], the Nelder–Mead simplex method [6], Powell's method [31], genetic algorithms [25, 59] or one of the various existing gradient ascent methods: steepest ascent, conjugate gradient, Levenberg–Marquardt, quasi-Newton methods, stochastic gradient [29, 60, 61]. The efficiency of a particular optimization method very much depends on the considered transformation space. Gradient-free methods such as Powell's or Nelder–Mead have been used extensively for low-dimensional (rigid or affine) image registration, but are computationally prohibitive in high-dimensional registration problems. Gradient-based methods, on the other hand, scale better with parameter dimension and are thus widely used for non-rigid image registration.

Apart from off-the-shelf methods, there exist optimization algorithms specifically developed for image registration. The generalized ICP algorithm discussed in Sections 7.2.1 and 7.3.1 may be seen as a relaxation method for intensity-based registration measures, and leads to an intuitive matching/registration iterative scheme [15, 62]. A closely related algorithm is block matching [63], where the matching step of ICP is constrained so that neighbouring voxels have neighbouring correspondents. In a different spirit, discrete optimization methods inspired by Markov random field segmentation models have been adapted to image registration [64, 65], and a current research trend in image registration is to apply general algorithms developed for inference in graphical models such as belief propagation [66, 67].

A difficulty common to many optimization methods is that they only guarantee convergence to a local maximum of the registration measure, and are thus sensitive to parameter initialization. In rigid/affine registration, convergence to a suboptimal registration solution is often easy to diagnose by visual inspection. Optimization may then be restarted from a manually determined initial transformation, provided that it is not too far off from the correct registration transformation. In non-rigid registration, convergence to a local maximum may be much more to diagnose, hence making the registration result highly dependent on the chosen optimization method.

7.4.1   Rigid Transformations

Rigid transformations (translation and rotation) are suitable for registration problems such as motion correction in longitudinal imaging studies or multimodal intra-patient registration, where the goal is to correct for different positioning of a subject with respect to the image acquisition devices. Three-dimensional rigid transformations are described by

$$T(\mathbf{x}) = \mathbf{R}\mathbf{x} + \mathbf{t},$$

where $\mathbf{t}$ is a 3×1 translation vector and $\mathbf{R}$ is a rotation matrix, that is it verifies the conditions $\mathbf{R}\mathbf{R}^\top = \mathbf{I}_3$ et $|\mathbf{R}| = 1$.

It should be noted that the space of rigid transformations is not a linear space, in the sense that a linear combination of rigid transformations is generally not a rigid transformation. Mathematically, rigid transformations define a Lie group of dimension six, meaning that every 3D rigid transformation can be uniquely described by six free parameters, specifically three translation parameters t_x, t_y, t_z, and three rotation parameters r_x, r_y, r_z, leading to the parametrization:

$$\theta = \begin{pmatrix} t_x & t_y & t_z & r_x & r_y & r_z \end{pmatrix}^\top.$$

There exist at least two effective ways to define the rotation parameters r_x, r_y, r_z, namely the Euler angles and the rotation vector [68]. Both representations involve three parameters. Another representation is the unitary quaternion, which involves four parameters related by a normality constraint and is therefore not well suited to unconstrained optimization.

7.4.2   Similarity Transformations

Similarity transformations generalize rigid transformations by including a global scaling factor,

$$T(\mathbf{x}) = s\mathbf{R}\mathbf{x} + \mathbf{t}, \qquad s > 0,$$

hence defining the simplest class of non-rigid transformations, which is a Lie group structure with dimension seven. Although similarity transformations do not preserve distances, they preserve angles and ratios of distances.

A simple way to force positive scales in practice is to parametrize the scale parameter by its exponential, that is $s = \exp(\theta_7)$ where θ_7 is the scale parameter.

7.4.3   Affine Transformations

General affine transformations are characterized by

$$T(\mathbf{x}) = \mathbf{A}\mathbf{x} + \mathbf{t},$$

where $\mathbf{A}$ is an arbitrary 3×3 matrix. Affine transformations generally do not preserve distances and angles, but preserve parallelism. They form a linear space of dimension 12.

A useful representation of affine transformations is obtained through the singular value decomposition,

$$\mathbf{A} = \mathbf{QSR}$$

where $\mathbf{R}$ and $\mathbf{Q}$ are rotation matrices (direct orthogonal matrices) and $\mathbf{S}$ is a diagonal matrix. If $\mathbf{S}$ differs from the identity matrix, the resulting affine transformation is not rigid and produces shearing effects. Note that it is often desired to force positive scaling factors in order to rule out reflections, which may not be plausible transformations.

Since any matrix $\mathbf{A}$ may be written under the above form, it is possible to parametrize affine transformations via

$$\theta = (t_x \quad t_y \quad t_z \quad r_x \quad r_y \quad r_z \quad s_x \quad s_y \quad s_z \quad q_x \quad q_y \quad q_z)^\mathsf{T},$$

where r_x, r_y, r_z and q_x, q_y, q_z are rotation parameters (rotation vectors or Euler angles), and s_x, s_y and s_z are scale parameters. As in the case of similarity transformations, we may force positive scaling factors by considering s_x, s_y and s_z as log-scales rather than actual scales.

7.4.4 Projective Transformations

Projective transformations are suitable for registration problems where a three-dimensional image is registered with a two-dimensional image acquired with a camera. They account for the perspective effects inherent to the two-dimensional image. There exist as many classes of projective transformations as camera models, the most common being the pinhole camera model. We refer the reader to [69] for an in-depth account of projective transformations.

7.4.5 Polyaffine Transformations

Polyaffine transformations formalize the notion of 'locally affine' transformations. In a simplistic version, they are defined as a spatially dependent linear combination of affine transformations,

$$T(\mathbf{x}) = \sum_{i=1}^{n} w_i(\mathbf{x})T_i(\mathbf{x}), \tag{7.20}$$

where, for each component i, $w_i(\mathbf{x})$ is a weighting function, for instance, a Gaussian kernel centred at some control point $\mathbf{c}_i$, and T_i is an affine transformation:

$$T_i(\mathbf{x}) = \mathbf{A}_i\mathbf{x} + \mathbf{t}_i.$$

Since there are n components, T is described by $12n$ free parameters.

This definition of polyaffine transformations, however, has a serious drawback: the transformation T as defined in Equation (7.20) is not guaranteed to be invertible, implying that it may not preserve topological relationships between objects and may thus not be physically plausible despite its apparent simplicity. An elegant way to guarantee invertibility is to modify the above definition according to [70]. Consider the principal logarithm of each affine transformation T_i, which is the matrix logarithm of the 4×4 matrix representing T_i in homogeneous coordinates:

$$\log \begin{pmatrix} \mathbf{A}_i & \mathbf{t}_i \\ 0 & 1 \end{pmatrix} = \begin{pmatrix} \mathbf{L}_i & \mathbf{v}_i \\ 0 & 0 \end{pmatrix}.$$

The idea is then to define the polyaffine transformation as the flow associated with the following ordinary differential equation (ODE):

$$\dot{\mathbf{x}} = \sum_{i=1}^{n} w_i(\mathbf{x})(\mathbf{v}_i + \mathbf{L}_i\mathbf{x}). \tag{7.21}$$

In other words, $T(\mathbf{x})$ is the position at time $t = 1$ of a particle with position $\mathbf{x}$ at time $t = 0$ submitted to the above ODE. The flow has the remarkable property to be a diffeomorphism (a smooth and smoothly invertible transformation), which follows from a general mathematical argument on non-autonomous differential equations [71]. The intuition behind this property is that the composition of many infinitesimal smooth displacements is smooth and invertible, unlike their sum.

In practice, $T(\mathbf{x})$ may be computed numerically using any ODE integration method. Note that in the case of a single affine component, $T_1(\mathbf{x})$, and provided that the weighting function is constant, $w_1(\mathbf{x}) \equiv 1$, the flow is trivially given by $T(\mathbf{x}) = T_1(\mathbf{x})$, meaning that the polyaffine transformation associated with a single affine transformation is that affine transformation itself, which is an expected consistency property.

7.4.6   Free-Form Transformations: 'Small Deformation' Model

Let us consider maximizing a registration measure $\tilde{S}(T)$ over the whole space of spatial transformations $T : \mathbb{R}^3 \to \mathbb{R}^3$. In this case, the transformation space is infinite dimensional, and it is easy to realize that the problem of maximizing $\tilde{S}(T)$ over such a huge space is ill-posed. The reason is that two distinct transformations that coincide on the source image grid have to yield the same registration measure, and thus the optimization problem cannot have a unique solution.

A classical way to work around ill-posedness is to add a regularization term to the objective function so as to rank solutions intrinsically. This leads to consider the following penalized maximization problem as in classical *optical flow* [72]:

$$\max_T [\tilde{S}(T) - \lambda U(T)] \tag{7.22}$$

where $U(T)$ is a regularization energy or stabilizer, and λ is positive constant. Quadratic stabilizers have long been studied for real-valued multivariate functions [73, 74] and can easily be generalized to vector-valued functions (such as spatial transformations) by summing up coordinate-wise energies [75], leading to *separable* stabilizers. Stabilizers are readily defined in the Fourier domain as their role is essentially to penalize 'high frequency' functions,

$$U(T) = \frac{1}{2} \sum_{i=1}^{3} \int_{\mathbb{R}^3} \frac{|\hat{T}^i(\mathbf{s})|^2}{\hat{G}(\mathbf{s})} \, d\mathbf{s}$$

where $\hat{T}^i(\mathbf{s})$ denotes the Fourier transform of the i-th coordinate of T and $\hat{G}(\mathbf{s})$ is a positive-valued function that vanishes as $\|\mathbf{s}\|$ goes to infinity. It is customary to choose $\hat{G}$ as a symmetric function, so that its inverse Fourier transform $G(\mathbf{x})$ is real-valued and symmetric. $G(\mathbf{x})$ plays the role of a *Green function* with respect to the stabilizer.

A necessary condition for T to maximize Equation (7.22) is that the gradient vanishes, yielding

$$\frac{\partial \tilde{S}}{\partial T} - \lambda \frac{\partial U}{\partial T} = \sum_k \frac{\partial \tilde{S}}{\partial T_k} \delta(\mathbf{x} - \mathbf{x}_k) - \lambda \frac{\partial U}{\partial T} = 0,$$

where we denote $T_k \equiv T(\mathbf{x}_k)$, hence $\partial \tilde{S}/\partial T_k$ is a 3×1 vector representing the gradient of $\tilde{S}$ with respect to a single voxel displacement. It follows that the optimal T verifies the following implicit equation:

$$T(\mathbf{x}) = T_0(\mathbf{x}) + \sum_k G(\mathbf{x} - \mathbf{x}_k)\, \mathbf{f}_k, \qquad \text{with} \quad \mathbf{f}_k \equiv \frac{1}{\lambda}\frac{\partial S}{\partial T_k}\bigg|_{T(\mathbf{x}_k)}, \qquad (7.23)$$

where T_0 is in the null space $\mathcal{N}$ of $U(T)$, that is it verifies $dU(T_0) = 0$. It is an implicit equation because the 'forces' $\mathbf{f}_k$ depend on T. For a Gaussian stabilizer, $\mathcal{N} = \{0\}$, therefore $T_0 = 0$. For other usual stabilizers, $\mathcal{N}$ is typically the space of affine transformations or a space of higher degree multivariate polynomials.

Equation (7.23) essentially tells us that the maximizer of Equation (7.22) is a multivariate spline with kernel function $G(\mathbf{x})$ and control points at every source grid voxel (note that, in order to save computation time and memory load in practice, it is customary to approximate this expression using a subset of control points $\mathbf{x}_k$). Equation (7.23) thus provides a finite-dimensional parametric representation of the solution, stating that the infinite-dimensional optimization problem of Equation (7.22) is equivalent to maximizing $\tilde{S}(T) - \lambda U(T)$ with respect to both the 'forces' $\mathbf{f}_k$ and the unknown parameters of T_0. This is a case where the transformation model is coupled with the variational problem associated with image registration; should another stabilizer be considered, the kernel G and thus the transformation model would change accordingly.

In this variational framework, the optimal transformation turns out to be a linear combination of elementary deformations. Whatever be the chosen basis (e.g. Gaussian, tensor B-spline, thin plate), such a linear representation is not guaranteed to yield an invertible transformation unless each component is assigned a small weight. In mathematical terms, the transformation may not be a diffeomorphism, that is, a smooth and smoothly invertible function. In simple words, it may either underestimate deformations, resulting in poor overlap between homologous structures, or produce non-invertible deformations that break topology. This is the same problem as that already encountered in Section 7.4.5 with polyaffine transformations.

Therefore, the penalized registration approach stated in Equation (7.22), which is akin to classical optical flow methods from computer vision [72], can only recover 'small' deformations and is inappropriate for registration problems that necessitate 'large' deformations. A typical large deformation problem is inter-subject brain image warping due to the high anatomical shape variability of the brain across individuals [3].

7.4.7 Free-Form Transformations: 'Large Deformation' Models

In the late 1990s, researchers figured out an alternative parametrization of free-form deformations that guarantees the diffeomorphic property. The core idea is to describe the transformation as the composition of many infinitesimal smooth displacements [71, 77]. This means that a linear parametrization can be used to represent not

Table 7.2 Some usual 3D transformation stabilizers with associated differential forms and kernel functions. Note that the kernel function corresponding to linear elasticity is an approximation taken from Ref. [76] where $\alpha = (\lambda + \mu)/(3\lambda + 7\mu)$

Stabilizer	Differential form dU	Kernel function $G(\mathbf{x})$
Thin plate	Δ^2	$\|\mathbf{x}\|$
Volume	$-\Delta^3$	$\|\mathbf{x}\|^3$
Gaussian	$\sum_{m=0}^{\infty} \frac{(-1)^m}{m!a^m} \Delta^m$	$e^{-a\|\mathbf{x}\|^2}$
Linear elasticity	$-\mu\Delta - (\lambda + \mu)\nabla(\nabla.)$	$\|\mathbf{x}\| \, I_3 - \alpha \frac{\mathbf{x}\mathbf{x}^t}{\|\mathbf{x}\|}$
Tensor B-spline	$(-1)^m \partial_x^{2m} \partial_y^{2m} \partial_z^{2m}$	$\beta_3^{2m-1}(\mathbf{x})$

the transformation itself but something analogous to a velocity field. The actual deformations are the displacements of particles submitted to the velocity field, yielding a typically non-linear evolution (Table 7.2).

7.4.7.1 Parametric Velocity

A simple implementation of this idea is to parametrize the velocity field, rather than the transformation itself, as a multivariate spline:

$$V(\mathbf{x}, t) = \sum_k G(\mathbf{x} - \mathbf{x}_k) \, \mathbf{f}_k,$$

where the $\mathbf{f}_k$'s are unknown vectors to be optimized and G is a Gaussian kernel (for the sake of clarity, we will restrict ourselves to Gaussian splines throughout the sequel). The velocity field is the convolution by a Gaussian kernel of a discrete 'force field', which is a vector-valued image with same grid dimension as the source image and takes on values $\mathbf{f}_1, \mathbf{f}_2, \ldots$ at positions $\mathbf{x}_1, \mathbf{x}_2, \ldots$, respectively. Hence, the velocity field is a smooth vector-valued function densely defined in space. The transformation T, which remains our quantity of interest, is then defined as the flow induced by the velocity field V:

$$\frac{d\phi}{dt} = V(\phi_t), \qquad \phi_0 = \text{Id}, \qquad T = \phi_1.$$

Therefore, computing $T(\mathbf{x})$ at any point $\mathbf{x}$ depending on the forces $\mathbf{f}_1, \mathbf{f}_2, \ldots$ requires integrating an ODE, which can be done using numerical methods. Note that, in this model, the velocity field itself is constant over time, which makes things relatively simple but is not required to ensure that the resulting T is a diffeomorphism.

Using the above parametrization for T, we may consider optimizing $\tilde{S}(T)$ or a penalized version of it, with respect to the force field using, for instance, a gradient ascent technique [61].

7.4.7.2 Geodesic Shooting

Another way to guarantee diffeomorphic registration is to modify the penalized registration problem of Equation (7.22) by another optimization problem where the variable to be optimized is a time-dependent velocity field V, that is a mapping from $\mathbb{R}^4$ to $\mathbb{R}^3$, leading to

$$\max_{V} \left[\tilde{S}(T) - \lambda \int_0^1 U(V_t)dt \right], \qquad (7.24)$$

where, again, T is determined by the velocity field V as the flow associated with V:

$$\frac{d\phi_t}{dt} = V_t(\phi_t), \qquad \phi_0 = \text{Id}, \quad T = \phi_1.$$

Note that, in this case, the velocity field is allowed to change over time. The only difference between the optimization problems of Equation (7.24) and Equation (7.22) lies in the respective energy terms used to penalize transformations *a priori*. The energy in Equation (7.24) may be interpreted as the 'kinetic energy' of an evolving fluid material. The method was coined 'geodesic shooting' [71] because the trajectory of a particle that minimizes its kinetic energy is the shortest path, or geodesic, between the endpoints.

In practice, the ODE is discretized using, for example, a fixed time step size ϵ, yielding an expression of T as a finite sequence of $N = 1/\epsilon$ compositions:

$$\phi_{n+1}(\mathbf{x}) = (\text{Id} + \epsilon V_n) \circ \phi_n(\mathbf{x}), \quad n = 0, 1, \dots, N-1, \qquad T = \phi_N.$$

The problem then reduces to finding the sequence of velocity fields $V_0, V_1, \dots, V_N$ that minimizes a variant of Equation (7.24) where the time integral is replaced by a discrete sum. Let us denote $\varphi_n = (\text{Id} + \epsilon V_N) \circ (\text{Id} + \epsilon V_{N-1}) \circ \dots \circ (\text{Id} + \epsilon V_{n+1})$ the total flow from time step $n+1$ to last. We have $T = \varphi_n \circ \phi_n$ and, equivalently, $\varphi_n = T \circ \phi_n^{-1}$. Differentiating Equation (7.24) with respect to the velocity field V_n at time step n, we get the necessary optimality condition:

$$V_n(\mathbf{x}) = \sum_k G(\mathbf{x} - \phi_{n-1}(\mathbf{x}_k)) \, d\varphi_n(\phi_n(\mathbf{x}_k))^{\top} \mathbf{f}_k, \qquad (7.25)$$

with $\mathbf{f}_k$ as in Equation (7.23), and where $d\varphi_n(\phi_n(\mathbf{x}_k))$ denotes the Jacobian matrix of φ_n computed at $\phi_n(\mathbf{x}_k)$. This expresses the velocity field at each time as a multivariate spline. However, this is an implicit equation because the right-hand side depends on the whole sequence of velocity fields $V_1, \dots, V_N$. Iterative numerical schemes can be used [3, 77–79] to solve for Equation (7.25), hence providing a numerical solution to Equation (7.24).

7.4.7.3 Demons Algorithm

A related, yet arguably simpler method to obtain a diffeomorphic transformation is the so-called 'Demons' algorithm [80–84] coined after Maxwell's demons because of an analogy with diffusing processes in thermodynamics. The basic idea is to iteratively increase the registration measure $\tilde{S}(T)$ by composing small smooth displacements, thereby producing a diffeomorphic transformation at each iteration owing to the same fundamental property as for flows of non-autonomous differential equations [71], which underpins the large deformation models presented in Sections 7.4.7.1 and 7.4.7.2.

Contrary to the parametric velocity approach of Section 7.4.7.1, the Demons method computes a sequence of *non-parametric* transformations $T_0, T_1, \ldots$ so that $\tilde{S}(T_n)$ increases at each iteration. Such a sequence of transformations may, at some point, come close to a reasonable registration solution. However, the process should *not* be iterated until convergence, because the maximizer of $\tilde{S}(T)$ over the whole space of diffeomorphisms is unlikely to be a realistic transformation. In other words, the Demons algorithm iterated for too long is not robust to image noise and tends to overfit the data. Hence, the number of iterations plays the role of a regularization parameter in the Demons. This is in contrast with the geodesic shooting method outlined in Section 7.4.7.2 where regularization is achieved by penalizing the registration measure using a 'kinematic' energy.

Typically, the initial transformation T_0 is the affine transformation that maximizes $\tilde{S}(T)$ and is computed using an independent optimization method. The goal at each iteration n is to compute a smooth incremental deformation field U so that the transformation update obtained by right-composing U with the current transformation,

$$T_{n+1} = T_n \circ U,$$

verifies $\tilde{S}(T_{n+1}) > \tilde{S}(T_n)$. If T_n is a diffeomorphism, so will be the update T_{n+1}, provided that U is smooth and 'close enough' to the identity transformation.

To obtain a suitable U, one may perform a gradient ascent iteration on $\tilde{S}(T_n \circ U)$ and smooth the result using some linear filter. Consider, for instance, a Levenberg–Marquardt iteration [85] and a Gaussian filter:

$$U(\mathbf{x}) = \mathbf{x} + \sum_k G(\mathbf{x} - \mathbf{x}_k)[\lambda_k I_3 + \mathbf{H}_n(\mathbf{x}_k)]^{-1}\mathbf{g}_n(\mathbf{x}_k), \tag{7.26}$$

where $\mathbf{g}_n(\mathbf{x}_k)$ and $\mathbf{H}_n(\mathbf{x}_k)$, respectively, denote estimates of the gradient and Hessian of $U \mapsto \tilde{S}(T_n \circ U)$ computed at $U = \text{Id}$, and the λ_k's are positive damping factors to be tuned. This is equivalent to convolving a discrete displacement field with a Gaussian filter, that is $U = \text{Id} + G \star D$ with

$$D(\mathbf{x}_k) \equiv \mathbf{d}_k = [\lambda_k I_3 + \mathbf{H}_n(\mathbf{x}_k)]^{-1}\mathbf{g}_n(\mathbf{x}_k). \tag{7.27}$$

In the case of the SSD measure (see Section 7.3.3), Equation (7.27) reads

$$\mathbf{d}_k = \frac{i_k - j_k^{\downarrow}}{\lambda_k + \|\nabla J_n(\mathbf{x}_k)\|^2}\nabla J_n(\mathbf{x}_k),$$

where $j_k^\downarrow$ and ∇J_n, respectively, denote the intensity and gradient of the resampled image $J_n \equiv J \circ T_n$ at location $\mathbf{x}_k$, and the Hessian is approximated via

$$\mathbf{H}_n(\mathbf{x}_k) \approx \nabla J_n(\mathbf{x}_k) \nabla J_n(\mathbf{x}_k)^\top.$$

If we choose $\lambda_k = (i_k - j_k^\downarrow)^2$, the unsmoothed displacements are bounded by half a voxel in the sense that $\|\mathbf{d}_k\| \le 0.5$, for any voxel index k, which still holds after smoothing and thus satisfies the requirement for small residual deformations. This is easily seen using the general inequality $2xy \le (x^2 + y^2)$, $\forall (x, y) \in \mathbb{R}^2$. There is no strict warranty that U as defined in Equation (7.26) verifies $\tilde{S}(T_n \circ U) > \tilde{S}(T_n)$. However, the discrete displacements obtained by the Levenberg–Marquardt iteration are likely to increase the registration measure; thus, by a continuity argument, we expect to have $\tilde{S}(T_n \circ (\text{Id} + G \star D)) > \tilde{S}(T_n)$, provided that the amount of smoothing is moderate. This can be tested, and damping factors can be increased until the test passes.

7.5 Conclusion

In this chapter, we reviewed some of the most widely used image registration techniques used in medical image analysis, emphasizing two important specificities of this field: (1) the need for registration measures that reflect complex intensity relationships between images acquired from different modalities or different acquisition protocols; (2) the need for topology-preserving (diffeomorphic) large deformation models in registration problems that involve images from different subjects. These two aspects were thoroughly discussed, respectively, in Section 7.3 and Section 7.4 from an algorithmic viewpoint. Mutual information and other pseudo-likelihood registration measures presented in Section 7.3 were motivated by their potential to cope with multimodal image registration and shown to generalize classical correlation measures used in computer vision which, in practice, are restricted to monomodal image registration problems. Furthermore, Section 7.4 discussed the parametrization of diffeomorphic transformations and presented several numerical schemes for the associated optimization of a registration measure.

7.6 Exercise

Consider the problem of warping two brain images from different subjects acquired from different modalities, namely T1-weighted and T2-weighted magnetic resonances, as depicted in Figure 7.10. The goal of this exercise is to implement the Demons algorithm (see Section 7.4.7.3) to non-rigidly register the T2-weighted image with the T1-weighted image using a pseudo-log-likelihood measure of the form defined in Section 7.3.5 as an objective function. We consider the T1-weighted image as the source image I and the T2-weighted image as the floating image J, hence applying deformations to the T2-weighted image.

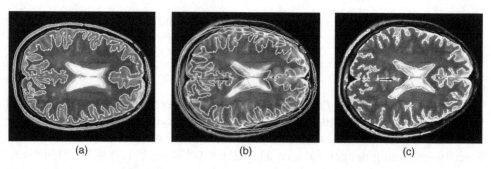

Figure 7.10 Profiles of three variants of the mutual information measure against translations with constant direction. (a) T1-weighted magnetic resonance image with Canny–Deriche contours overlaid, (b) T2-weighted image from an other subject with T1-weighted image contours overlaid and (c) objective registered T2-weighted image

The pseudo-log-likelihood measure will be instantiated by assuming that the T2 intensity is normally distributed *conditionally* on the T1 intensity, meaning that the distribution $p(i, j)$ involved in Equation (7.12) reads

$$p(i, j) = p(j|i)p(i), \qquad \text{with} \quad p(j|i) = N(f(i), \sigma).$$

where f and σ are, respectively, an intensity mapping and a standard deviation parameter to be estimated. We further assume that $\pi(i)$ and $p(j)$ in Equation (7.12) are such that

$$\pi(i) = p(i), \qquad \pi(j) = N(\mu_0, \sigma_0),$$

where μ_0 and σ_0 are additional parameters.

1. Prove that the pseudo-log-likelihood measure reads as follows regardless of $p(i)$:

$$\bar{S}(T, f, \sigma, \mu_0, \sigma_0) = -n_T \log \frac{\sigma}{\sigma_0} - \frac{1}{2} \sum_{k \in \mathcal{A}^\downarrow} \left[\frac{\left(j_k^\downarrow - f(i_k) \right)^2}{\sigma^2} - \frac{\left(j_k^\downarrow - \mu_0 \right)^2}{\sigma_0^2} \right].$$

2. Argue that if μ_0 and σ_0 are considered known parameters, then in the limit of large σ_0 the above expression may be approximated (up to a large constant) by

$$\bar{S}(T, f, \sigma) \approx -n_T \log \sigma - \frac{1}{2\sigma^2} \sum_{k \in \mathcal{A}^\downarrow} \left[j_k^\downarrow - f(i_k) \right]^2 + \text{constant}.$$

3. Assuming that f is piecewise constant on intervals of the form $[i - \frac{1}{2}, i + \frac{1}{2}[$, where i spans the set of integers, show that, at fixed T, $\bar{S}(T, f, \sigma)$ is maximized by

$$\hat{f}(i) = \frac{1}{\# \mathcal{A}_i^\downarrow} \sum_{k \in \mathcal{A}_i^\downarrow} j_k^\downarrow, \qquad \hat{\sigma}^2 = \frac{1}{n_T} \sum_{k \in \mathcal{A}^\downarrow} \left[j_k^\downarrow - \bar{f}(i_k) \right]^2, \qquad (7.28)$$

where $\mathcal{A}_i^{\downarrow}$ is the subset of source voxels with intensity in $[i - \frac{1}{2}, i + \frac{1}{2}[$, that is $\mathcal{A}_i^{\downarrow} = \mathcal{A}^{\downarrow} \cap \{k, i - \frac{1}{2} \le i_k < i + \frac{1}{2}\}$.

4. Consider performing a Demons iteration at fixed f and σ to update the displacement field applied to image J. Using the same Hessian approximation as in Section 7.4.7.3 for the SSD measure, and the same argument to limit all incremental displacements to half a voxel, show that the raw incremental displacement field (before spatial smoothing) as defined in Equation (7.27) takes the form:

$$\mathbf{d}_k = \mathbf{x}_k + \frac{f(i_k) - j_k^{\downarrow}}{[f(i_k) - j_k^{\downarrow}]^2 + \|\nabla J_n(\mathbf{x}_k)\|^2} \nabla J_n(\mathbf{x}_k). \qquad (7.29)$$

5. Implement an algorithm that alternates Demons steps for the estimation of spatial displacements and maximizations of $\bar{S}(T, f, \sigma)$ w.r.t. (f, σ).

7.6.1 Implementation Guidelines

This algorithm boils down to a simple modification of the standard SSD-based Demons algorithm where displacement field updates are interleaved with source image intensity correction steps. It can be implemented using the following iterative procedure given in pseudo-code with a fixed number of iterations N:

Algorithm sketch

Initialization: $T =$ Id.

```
For iter= 1,...,N:
```
- Resample the floating image according to the current transformation estimate:
```
Jn = apply_transformation(T, J)
```
- Compute the image gradient ∇J_n:
```
dJn = image_gradient(Jn)
```
- Correct the source image intensity:
```
In = correct_intensity(I, Jn)
```
- Compute the raw incremental displacement field:
```
D = compute_raw_displa-
cements(In, Jn, dJn)
```
- Compute the incremental transformation U by smoothing the incremental displacement field using a Gaussian kernel:
```
U = incremental_transformation(D)
```
- Update transformation T by right-composition with the smoothed incremental displacement field, $T \leftarrow T \circ U$:
```
T = compose_transformations(T,U)
```

The functions involved in the above loop constitute the building blocks of the algorithm, and can be implemented using the following guidelines:

Function `apply_transformation`: Resample the floating image *J* to the same dimensions as the source image *I* according to the current spatial transformation *T*, represented by a three-component vector-valued image. This procedure involves intensity interpolation, for which a good tradeoff between computation time and accuracy is generally achieved using trilinear interpolation. Voxels that transform outside the image boundary may be assigned a zero intensity.

Function `image_gradient`: Compute the gradient of an image, resulting in a three-component vector-valued image. This may be implemented in practice using either finite differences or Gaussian derivative kernels.

Function `correct_intensity`: Depending on both the source image *I* and the resampled floating image J_n, output an intensity-corrected version of *I*, that is $f(I)$, by applying Equation (7.28) at each voxel.

Function `compute_raw_displacements`: Output a displacement field, represented by a vector-valued image, by applying Equation (7.29) at each voxel depending on the corrected source image, the currently resampled floating image and its gradient.

Function `incremental_transformation`: Output a vector-valued image $U = \mathrm{Id} + G \star D$ of transformed locations by smoothing the incremental displacement field using a Gaussian kernel. We may typically choose an isotropic kernel with 1 voxel standard deviation.

Function `compose_transformations`: Compute $T(U(\mathbf{x}_k))$ for each location $\mathbf{x}_k$ in the source image. Given that *T* is represented by a vector-valued image, this step involves interpolating values of *T*, and again trilinear interpolation can safely be used.

References

[1] Brown, L.G. (1992) A survey of image registration techniques. *ACM Computing Surveys*, **24** (4), 325–376.

[2] Toga, A.W. (2002) *Brain Warping: The Methods*, 2nd edn, Academic Press.

[3] Klein, A., Andersson, J., Ardekani, B.A. *et al.* (2008) Evaluation of 14 nonlinear deformation algorithms applied to human brain. *NeuroImage*, **46** (3), 786–802.

[4] van den Elsen, P., Pol, E. and Viergever, M. (1993) Medical image matching - a review with classification. *IEEE Engineering in Medicine and Biology*, **12** (4), 26–39.

[5] Lavallée, S. (1995) Registration for computer integrated surgery: methodology, state of the art, in *Computer Integrated Surgery* (eds R. Taylor, S. Lavallee, G. Burdea and R. Moesges), MIT Press, pp. 77–97.

[6] Maintz, J.B.A. and Viergever, M.A. (1998) A survey of medical image registration. *Medical Image Analysis*, **2** (1), 1–36.

[7] Hill, D., Batchelor, P., Holden, M. and Hawkes, D. (2001) Medical image registration. *Physics in Medicine and Biology*, **46**, 1–45.

[8] Zitova, B. and Flusser, J. (2003) Image registration methods: a survey. *Image and Vision Computing*, **21** (11), 977–1000.

[9] Pennec, X. (1998) Toward a generic framework for recognition based on uncertain geometric features. *Videre: Journal of Computer Vision Research*, **1** (2), 58–87.

[10] Rangarajan, A., Chui, H. and Bookstein, F. (1997) The softassign procrustes matching algorithm, *International Conference on Information Processing in Medical Imaging (IPMI)*, LNCS, Vol. **3**, Springer-Verlag, pp. 29–42.

[11] Rangarajan, A., Chui, H. and Duncan, J. (1999) Rigid point feature registration using mutual information. *Medical Image Analysis*, **3** (4), 425–440.

[12] Granger, S., Pennec, X. and Roche, A. (2001) Rigid point-surface registration using an EM variant of ICP for computed guided oral implantology, in *Medical Image Computing and Computer-Assisted Intervention (MICCAI'01)*, *Lecture Notes in Computer Science*, Vol. **2208** (eds W. Niessen and M. Viergever), Springer-Verlag, Utrecht, NL, pp. 752–761.

[13] Besl, P. and McKay, N. (1992) A method for registration of 3-D shapes. *IEEE Transactions on Pattern Analysis and Machine Intelligence*, **14** (2), 239–256.

[14] Zhang, Z. (1994) Iterative point matching for registration of free-form curves and surfaces. *International Journal of Computer Vision*, **13** (2), 119–152.

[15] Feldmar, J., Declerck, J., Malandain, G. and Ayache, N. (1997) Extension of the ICP algorithm to non-rigid intensity-based registration of 3D volumes. *Computer Vision and Image Understanding*, **66** (2), 193–206.

[16] Eggert, D., Lorusso, A. and Fisher, R. (1997) Estimating 3D rigid body transformations: a comparison of four major algorithms. *Machine Vision Applications, Special Issue on Performance Characteristics of Vision Algorithms*, **9** (5/6), 272–290.

[17] Barnea, D. and Silverman, H. (1972) A class of algorithms for fast digital image registration. *IEEE Transactions on Computers*, **21** (2), 179–186.

[18] McGillem, C.D. and Svedlow, M. (1976) Image registration error variance as a measure of overlay quality. *IEEE Transactions on Geoscience Electronics*, **4** (1), 44–49.

[19] Svedlow, M., McGillem, C. and Anuta, P. (1978) Image registration: similarity measure and processing method comparisons. *AeroSys*, **14** (1), 141–149.

[20] Woods, R.P., Mazziotta, J.C. and Cherry, S.R. (1993) MRI-PET registration with automated algorithm. *Journal of Computer Assisted Tomography*, **17** (4), 536–546.

[21] Ardekani, B.A., Braun, M., Hutton, B.F. *et al.* (1995) A fully automatic multimodality image registration algorithm. *Journal of Computer Assisted Tomography*, **19** (4), 615–623.

[22] Alpert, N.M., Berdichevsky, D., Levin, Z. *et al.* (1996) Improved methods for image registration. *NeuroImage*, **3**, 10–18.

[23] Nikou, C., Heitz, F. and Armspach, J. (1999) Robust voxel similarity metrics for the registration of dissimilar single and multimodal images. *Pattern Recognition*, **32** (8), 1351–1368.

[24] Roche, A., Malandain, G., Pennec, X. and Ayache, N. (1998) The correlation ratio as a new similarity measure for multimodal image registration, *Medical Image Computing and Computer-Assisted Intervention (MICCAI'98)*, Lecture Notes in Computer Science, Vol. **1496**, Springer-Verlag, Cambridge, MA, pp. 1115–1124.

[25] Hill, D., Studholme, C. and Hawkes, D. (1994) Voxel similarity measures for automated image registration, *Visualization in Biomedical Computing*, SPIE Proceedings, Vol. **2359**, SPIE Press, pp. 205–216.

[26] Collignon, A., Vandermeulen, D., Suetens, P. and Marchal, G. (1995) 3D multi-modality medical image registration using feature space clustering, In *Computer Vision, Virtual Reality and Robotics in Medicine (CVRMed)*, LNCS, Vol. **905** (ed. N. Ayache), Springer-Verlag, pp. 195–204.

[27] Studholme, C., Hill, D.L.G. and Hawkes, D.J. (1995) Multiresolution voxel similarity measures for MR-PET registration, in *International Conference on Information Processing in Medical Imaging (IPMI)* (eds Y. Bizais and C. Barillot), Kluwer Academic Publishers, pp. 287–298.

[28] Collignon, A., Maes, F., Delaere, D. *et al.* (1995) Automated multi-modality image registration based on information theory, in *International Conference on Information Processing in Medical Imaging (IPMI)* (eds Y. Bizais and C. Barillot), Kluwer Academic Publishers, pp. 263–274.

[29] Viola, P. (1995) Alignment by maximization of mutual information. PhD thesis. Massachusetts Institute of Technology.

[30] Wells, W.M., Viola, P., Atsumi, H. and Nakajima, S. (1996) Multi-modal volume registration by maximization of mutual information. *Medical Image Analysis*, **1** (1), 35–51.

[31] Maes, F., Collignon, A., Vandermeulen, D. *et al.* (1997) Multimodality image registration by maximization of mutual information. *IEEE Transactions on Medical Imaging*, **16** (2), 187–198.

[32] Studholme, C., Hill, D.L.G. and Hawkes, D.J. (1998) An overlap invariant entropy measure of 3D medical image alignment. *Pattern Recognition*, **1** (32), 71–86.

[33] West, J. *et al.* (1997) Comparison and evaluation of retrospective intermodality brain image registration techniques. *Journal of Computer Assisted Tomography*, **21**, 554–566.

[34] Costa, W.L.S., Haynor, D.R., Haralick, R.M. *et al.* (1993) A maximum-likelihood approach to PET emission/attenuation image registration. *IEEE Nuclear Science Symposium and Medical Imaging Conference, Vol. 2*, IEEE, San Francisco, CA, pp. 1139–1143.

[35] Leventon, M.E. and Grimson, W.E.L. (1998) Multi-modal volume registration using joint intensity distributions, *Medical Image Computing and Computer-Assisted Intervention (MICCAI), LNCS*, Vol. **1496**, Springer-Verlag, Cambridge, MA, pp. 1057–1066.

[36] Bansal, R., Staib, L.H., Chen, Z. *et al.* (1998) A novel approach for the registration of 2D portal and 3D CT images for treatment setup verification in radiotherapy, *Medical Image Computing and Computer-Assisted Intervention (MICCAI)*, LNCS, Vol. **1496**, Springer-Verlag, Cambridge, MA, pp. 1075–1086.

[37] Roche, A., Malandain, G. and Ayache, N. (2000) Unifying maximum likelihood approaches in medical image registration. *International Journal of Imaging Systems and Technology: Special Issue on 3D Imaging*, Vol. 11, pp. 71–80.

[38] Zhu, Y.-M. and Cochoff, S.M. (2002) Likelihood maximization approach to image registration. *IEEE Transactions on Image Processing*, **11** (12), 1417–1426.

[39] Zöllei, L., Fisher, J.W. and Wells, W.M. (2003) A unified statistical and information theoretic framework for multi-modal image registration, *Information Processing in Medical Imaging*, Springer-Verlag, pp. 366–377.

[40] Unser, M. (1999) Splines: a perfect fit for signal/image processing. *IEEE Signal Processing Magazine*, **16** (6), 22–38.

[41] Unser, M., Aldroubi, A. and Eden, M. (1993) B-spline signal processing: Part I-theory. *IEEE Transactions on Signal Processing*, **30** (2), 821–833.

[42] Unser, M., Aldroubi, A. and Eden, M. (1993) B-spline signal processing: Part II-efficient design and applications. *IEEE Transactions on Signal Processing*, **30** (2), 834–848.

[43] Grünwald, P. (2007) *The Minimum Description Length Principle*, MIT Press.

[44] Baggenstoss, P. (2003) The PDF projection theorem and the class-specific method. *IEEE Transactions on Signal Processing*, **51** (3), 672–685.

[45] Chung, A.C.S., Wells, W.M. III, Norbash, A. and Grimson, W.E.L. (2002) Multi-modal image registration by minimising Kullback-Leibler distance, *Medical Image Computing and Computer-Assisted Intervention (MICCAI), LNCS*, Vol. **2489**, Springer-Verlag, Tokyo, pp. 525–532.

[46] Cover, T.M. and Thomas, J.A. (1991) *Elements of Information Theory*, John Wiley & Sons, Inc.

[47] Pluim, J.P.W., Maintz, J.B.A. and Viergever, M.A. (2000) Interpolation artefacts in mutual information based image registration. *Computer Vision and Image Understanding*, **77** (2), 211–232.

[48] Tsao, J. (2003) Interpolation artifacts in multimodality image registration based on maximization of mutual information. *IEEE Transactions on Medical Imaging*, **22** (7), 854–864.

[49] Ji, J.X., Pan, H. and Liang, Z.-P. (2003) Further analysis of interpolation effects in mutual information-based image registration. *IEEE Transactions on Medical Imaging*, **22** (9), 1131–1140.

[50] Liu, C. (2003) An Example of Algorithm Mining: Covariance Adjustment to Accelerate EM and Gibbs, To appear in Development of Modern Statistics and Related Topics.

[51] Salvado, O. and Wilson, D.L. (2007) Removal of local and biased global maxima in intensity-based registration. *Medical Image Analysis*, **11** (2), 183–196.

[52] Ashburner, J. and Friston, K. (2005) Unified segmentation. *NeuroImage*, **26** (3), 839–851.

[53] Dempster, A.P., Laird, N.M. and Rubin, D.B. (1977) Maximum likelihood from incomplete data via the EM algorithm. *Journal of the Royal Statistical Society: Series B*, **39**, 1–38.

[54] Jenkinson, M. and Smith, S. (2001) Global optimisation for robust affine registration. *Medical Image Analysis*, **5** (2), 143–156.

[55] Sarrut, D. and Feschet, F. (1999) The partial intensity difference interpolation, in *International Conference on Imaging Science, Systems and Technology* (ed. H.R. Arabnia), CSREA Press, Las Vegas, NV.

[56] Viola, P. and Wells, W.M. (1997) Alignment by maximization of mutual information. *International Journal of Computer Vision*, **24** (2), 137–154.

[57] De Nigris, D., Collins, D. and Arbel, T. (2012) Multi-modal image registration based on gradient orientations of minimal uncertainty. *IEEE Transactions on Medical Imaging*, **31** (12), 2343–2354.

[58] Studholme, C., Hill, D.L.G. and Hawkes, D.J. (1996) Automated 3-D registration of MR and CT images of the head. *Medical Image Analysis*, **1** (2), 163–175.

[59] Kruggel, F. and Bartenstein, P. (1995) Automatical registration of brain volume datasets, in *International Conference on Information Processing in Medical Imaging (IPMI)* (eds Y. Bizais and C. Barillot), Kluwer Academic Publishers, pp. 389–390.

[60] Maes, F., Vandermeulen, D. and Suetens, P. (1999) Comparative evaluation of multiresolution optimization strategies for multimodality image registration by maximization of mutual information. *Medical Image Analysis*, **3** (4), 373–386.

[61] Ashburner, J. (2007) A fast diffeomorphic image registration algorithm. *NeuroImage*, **38** (1), 95–113.

[62] Cachier, P., Bardinet, E., Dormont, D. *et al.* (2003) Iconic feature based nonrigid registration: the pasha algorithm. *Computer Vision and Image Understanding*, **89** (2-3), 272–298, Special Issue on Nonrigid Registration.

[63] Ourselin, S., Roche, A., Prima, S. and Ayache, N. (2000) Block matching: a general framework to improve robustness of rigid registration of medical images, in *Medical Image Computing and Computer-Assisted Intervention (MICCAI'00)*, Lecture Notes in Computer Science, Vol. **1935** (eds S. Delp, A. Di Gioia and B. Jaramaz), Springer-Verlag, Pittsburgh, PA, pp. 557–566.

[64] Glocker, B., Komodakis, N., Tziritas, G. *et al.* (2008) Dense image registration through MRFs and efficient linear programming. *Medical Image Analysis*, **12** (6), 731–741.

[65] Zikic, D., Glocker, B., Kutter, O. *et al.* (2010) Linear intensity-based image registration by Markov random fields and discrete optimization. *Medical Image Analysis*, **14** (4), 550–562.

[66] Sun, J., Zheng, N.-N. and Shum, H.-Y. (2003) Stereo matching using belief propagation. *IEEE Transactions on Pattern Analysis and Machine Intelligence*, **25** (7), 787–800.

[67] Szeliski, R. (2006) Image alignment and stitching: a tutorial. *Foundations and Trends®; in Computer Graphics and Vision*, **2** (1), 1–104.

[68] Ayache, N. (1991) *Artificial Vision for Mobile Robots*, The MIT Press, Cambridge, MA.

[69] Faugeras, O. (1993) *Three-Dimensional Computer Vision*, The MIT Press.

[70] Arsigny, V., Commowick, O., Ayache, N. and Pennec, X. (2009) A fast and log-Euclidean polyaffine framework for locally linear registration. *Journal of Mathematical Imaging and Vision*, **33** (2), 222–238.

[71] Miller, M., Trouvé, A. and Younes, L. (2006) Geodesic shooting for computational anatomy. *Journal of Mathematical Imaging and Vision*, **24** (2), 209–228.

[72] Horn, B. and Schunck, B. (1981) Determining optical flow. *Artificial Intelligence*, **17**, 185–203.

[73] Wahba, G. (1990) *Spline Models for Observational Data*, Society for Industrial and Applied Mathematics.

[74] Girosi, F., Jones, M. and Poggio, T. (1995) Regularization theory and neural networks architectures. *Neural Computation*, **7**, 219–269.

[75] Terzopoulos, D. (1986) Regularization of inverse visual problems involving discontinuities. *IEEE Transactions on Pattern Analysis and Machine Intelligence*, **8** (4), 413–424.

[76] Davis, M.H., Khotanzad, A., Flamig, D.P. and Harms, S.E. (1997) A physics-based coordinate transformation for 3-D image matching. *IEEE Transactions on Medical Imaging*, **16** (3), 317–328.

[77] Joshi, S.C. and Miller, M.I. (2000) Landmark matching via large deformation diffeomorphisms. *IEEE Transactions on Image Processing*, **9** (8), 1357–1370.

[78] Beg, M.F., Miller, M.I., Trouvé, A. and Younes, L. (2005) Computing large deformation metric mappings via geodesic flows of diffeomorphisms. *International Journal of Computer Vision*, **61** (2), 139–157.

[79] Ashburner, J. and Friston, K. (2011) Diffeomorphic registration using geodesic shooting and Gauss-Newton optimisation. *NeuroImage*, **55** (3), 954–967.

[80] Thirion, J.-P. (1998) Image matching as a diffusion process: an analogy with Maxwell's demons. *Medical Image Analysis*, **2** (3), 243–260.

[81] Guimond, A., Roche, A., Ayache, N. and Meunier, J. (2001) Multimodal brain warping using the demons algorithm and adaptative intensity corrections. *IEEE Transactions on Medical Imaging*, **20** (1), 58–69.

[82] Chefd'hotel, C. Hermosillo, G. and Faugeras, O. (2002) Flows of diffeomorphisms for multimodal image registration. Proceedings of the IEEE International Symposium on Biomedical Imaging, pp. 753–756.

[83] Vercauteren, T., Pennec, X., Perchant, A. and Ayache, N. (2009) Diffeomorphic demons: efficient non-parametric image registration. *NeuroImage*, **45** (1), S61–S72.

[84] Cahill, N., Noble, J. and Hawkes, D. (2009) A demons algorithm for image registration with locally adaptive regularization, *Medical Image Computing and Computer-Assisted Intervention (MICCAI)*, *LNCS*, Springer, Vol. **5761**, pp. 574–581.

[85] Press, W.H., Flannery, B.P., Teukolsky, S.A. and Vetterling, W.T. (1992) *Numerical Recipes in C*, 2nd edn, Cambridge University Press.

8

Video Restoration Using Motion Information

8.1 Introduction

Video enhancement and restoration have always been important in applications related to analogue and digital recording of image sequences. Even the new advanced cameras for video acquisition may suffer from severe degradations due to either the imperfections of the hardware or the uncontrollable acquisition conditions such as in medical, surveillance and satellite imaging. Not only restoration provides mechanisms to improve the visual quality of the captured video, but it can improve indirectly the performance of subsequent tasks related to computer vision and machine learning.

Many of the historically significant records are either unavailable to their format or too fragile to survive any attempt at copy and playback. In many cases, the only surviving record can be found on a different medium such as videotape on which artefacts are placed in the image frames due to the multiple generations of copying or the transfers between different media. Furthermore, old film sequences are likely to have suffered degradations at the early stage of the acquisition mainly due to hardware and technological limitations. Examples are irregular frame rates and unstable exposure due to the imprecise shutter mechanisms. Also, films may suffer further damage due to environmental hazards such as humidity, dust and even improper storage. In conventional manual restoration process, the complexity and the associated cost are inappropriate for a large number of films since dedicated equipment is required that can target a limited range of artefacts and the expected time is not appropriate for large scale video data sets. Consequently, the preservation of motion pictures and video tapes recorded over the last century became an important application of video

Image, Video & 3D Data Registration: Medical, Satellite & Video Processing Applications with Quality Metrics, First Edition. Vasileios Argyriou, Jesus Martinez Del Rincon, Barbara Villarini and Alexis Roche. © 2015 John Wiley & Sons, Ltd. Published 2015 by John Wiley & Sons, Ltd.

enhancement and restoration. Both professional archivists and broadcasting companies are interested in preserving these fragile archives, aiming to achieve a visual quality that meets the standards of today.

Automatic restoration is the only feasible approach towards a successful exploitation of film and video archives. The viewing experience can be significantly enriched by improving the frame quality and by reducing the perceptual impact meeting the viewers' aesthetic expectations. The restoration process initially involves a transformation stage of the archived films or videos from their original form to a digital media. During the second stage of this process, all kinds of degradations are detected and removed from the digitized image sequences increasing their visual quality. Restoration also may improve the compression level of digital image sequences since noise is removed and the original special and temporal correlation is restored. Particularly, the suppression of such artefacts has vital implications of the efficiency of video coding standards such as MPEG-2 and MPEG-4. Finally, restored and enhanced video sequences are also likely to contribute more to efficient management of pictorial databases and data mining.

Both terms restoration and enhancement were used earlier, but it is important to understand the difference between them. Video restoration is the process of taking a corrupted or noisy image sequence and estimating the original image. While enhancement emphasizes features that make the video more pleasing to the observer, but not necessarily to produce realistic data from a scientific point of view. Also, another important difference between enhancement and restoration of video sequences is the amount of data to be processed. Therefore, these methods should have a manageable complexity and be either automatic or semi-automatic.

Restoration of films and videos arises as soon as the condition of the media starts to deteriorate causing visible loss of quality. There are many factors that can produce degradation such as irregular frame rate, unstable exposure, standards conversion, mechanical damage or poor storage introducing either unsteadiness or loss of resolution. Therefore, the techniques for video restoration that were developed allow intervention at the frame, pixel or subpixel level. As a consequence, the restoration tools and algorithms include motion estimation and registration techniques, statistical processing and other mathematical methods [1–5].

Today with the emergence of new multimedia techniques, restoration is regarded as a key element towards the successful exploitation of film and video archives. One of the main issues that increase the complexity of the restoration process is the fact that image sequences may be degraded by multiple artefacts. Therefore, a single method for restoring all of the artefacts simultaneously is not feasible. In practice, a sequential approach is followed in which artefacts are removed one by one. In general, the whole restoration process is divided into three main parts as it is illustrated in Figure 8.1.

As it was mentioned earlier in film and video archives, a large variety of artefacts are encountered; in most of them, the successful removal relies on registration algorithms exploiting the temporal relationships between successive frames. In this chapter, we discuss registration and motion estimation techniques utilized mainly in a range of

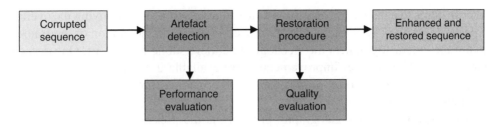

Figure 8.1   Processing steps in image sequence restoration

artefacts including video noise and grain, unsteadiness, flicker and dirt. The restoration mechanisms that are analysed are automatic, and the whole procedure could be applied in real time. The term real time indicates that the restoration and enhancement mechanisms could be applied when the video or film is broadcasted and without any prior knowledge about the type of the artefacts or their range.

Finally, it should be pointed out that an enormous amount of research over the past three decades was focused on image and video restoration. The research conducted on single static images can be applied also to image sequences assuming no temporal correlation. Of course, significant improvement can be obtained if inter-frame methods based on registration are utilized. In this chapter, we focus on analysing artefacts on image sequences and restoration techniques based mainly on registration and the exploitation of inter-frame information. Furthermore, metrics for performance evaluation and quality assessment are discussed.

8.2   History of Video and Film Restoration

The first device that enabled a person to watch moving images was Kinetoscope and was developed by the Edison company in 1891. The concept was similar to a photograph, but a cylinder was used to rotate the film (light-sensitive material) taking another picture. The reproduction was based on the same mechanism with the already processed pictures to run through a viewer creating the illusion of motion.

In 1895, the first Cinematograph device was developed by the Lumiere brothers, and the world's first public film screening was held in Paris lasting only 20 min in total. This device was able to function as a camera, projector and printer, all in one enshrining cinema as the art form of the twentieth century.

Full colour films started to be produced by 1906 when the principles of colour separation were introduced. During the first years of colour films, the cost was significantly high and the quality low, but after 1932 when the three-colour process was introduced it started to become a standard. Also, synchronized sound became available in the 1930s, and in the 1940s the quality improved and mechanisms to eliminate potential hazards were developed. The film industry faced a declining audience due to the rise of television in the early 1950s. Cinema started to use special effects, 3D formats and other features to enhance the quality and attract more people by improving the overall experience. In the 1970s, the quality improved focusing on the motion of the frames giving the cinematographers the freedom to be more subtle in their art. One decade later in the 1980s, further improvements were available in the area of colour, contrast and speed. Also, time code was embedded providing significant help during the production and the telecine transfers.

Nowadays, digital cinematography probably is the greatest leap since the introduction of sound due to the introduced digital production and post-production technologies. The films now include digital special effects, computer-generated imagery (CGI), digital image enhancement, animations, augmented reality and so on, moving the cinema experience beyond merely creating eye-catching effects and having a significant impact on the aesthetics, the production and the overall quality.

Broadcasting television signals was technically feasible around 1930 through a series of technical innovations both in Europe and in the United States. The concept of representing and converting light brightness to electrical impulses using photosensitive material was introduced in 1883 by Paul Nipkow. During his studies at the University of Berlin, the design and the principles of the Nipkow disc were presented, a disc capable of registering up to 4000 pixels per second. Due to the lack of the required hardware that period delayed further the development until 1925, when John Logie Baird developed an electromagnetic system capable of displaying a vertically scanned 30-line red and black image.

The first commercial television called televisor was available in the 1930s for home reception for about 18 lb. In 1928, a smaller device capable to display 30 lines and 10 frames per second (fps) was presented at the Berlin Radio Show by Denes von Mihaly. Vladimir Zworykin was the inventor of the first television based on electronic scanning around 1929. The cathode-ray tube (CRT) oscilloscope developed in 1897 by Karl Ferdinand Braun made possible Zworykin's designs, and the final device was introduced to the public in 1939. Also the period that interlaced scanning was independently developed. Later in the mid-1930s, outside broadcasts were possible using portable transmitters mounted on special vehicles. During the following decade, the innovations and the production were at their minimum level due to the World War, but in 1950s the number of television sets increased exponentially in the whole world.

In the 1980s and 1990s, the technological advantages started to occur at an even faster pace, resulting in the development of video recording devices allowing both the storage and the editing of the captured image sequences. At that stage, small variations existed in the standards, so with very few exceptions, films ran at the speed of 24 fps. NTSC video was running at 30 fps and PAL at 25 fps. Regarding the fields, the smallest temporal unit consisting of half of the scan line of a television frame, NTSC has 60 fields per second and PAL has 50 fields/s. A summary of the basic specifications is shown in Table 8.1.

An issue that is related also with video restoration is the standards conversion, since degradation may occur during that process. In more details, converting from film to PAL is more straightforward because the movie and the PAL frame rate are almost the

Table 8.1 Summary of film and video specifications

	Film	Video
Frame rate	24 frames per second (12 fps duplicated)	50 fields per second with 2 interlaced fields per frame result to 25 frames per second
Resolution	3000 lines per frame height (35 mm negative)	576 lines per frame height (PAL) 478 lines per frame height (NTSC)
Contrast ratio	1000:1 (negative) 100:1 (cinema screen)	20:1 (domestic receiver)
Aspect ratio	1.66 : 11.85 : 1	1.33 : 1(4 : 3) 1.77 : 1(16 : 9) HDTV, widescreen
Colour	Subtractive (secondary colours: yellow, magenta, cyan)	Additive (primary colours: red, green, blue)
Gamma	0.55 (colour negative) 3.0 (colour print)	1.0 (camera) 2.2 (display)

same (24 and 25 fps respectively). In order to synchronize them the movie is played 4% faster, and as a result a 1 h movie lasts 2 min and 24 s less of the original film. This procedure makes also the voices sound a half-tone higher, which even for audio quality it is considered as detrimental, it is unnoticeable for the majority of the viewers. It has been known that some people complained that the songs in movies such as Pink Floyd's The Wall are too fast and out of tune when they are transferred to PAL video. In the other case, converting to NTSC requires a different approach since the frame rate is higher at 30 fps. The solution in this case is based on the fact that five fields take $5/60$ s $= 1/12$ s, which is the time it takes for a film to show two frames. Therefore, the first film frame is shown for three consecutive NTSC fields, and the second film frame for two NTSC fields. As a result, the odd film frames are presented for $1/20$ s, while the even for $1/30$ s. This introduces motion judder especially during camera pans and slow motion, and due to the fact that the NTSC rate is 59.97 the error is only 1.8 s/h. In Figure 8.2, all these conversions are shown [1, 6, 7].

In the modern era, these problems are not significant anymore due to the digital cinema and the new standards that are utilized. Also, the quality of sound and image has been improved, and cable, satellite, high definition and interactive services became available. In 1998, the United Kingdom switched on the first digital television network in the world while Spain and Sweden followed in 1999. Nowadays, most of the countries in the world switched on the digital broadcasting, providing more facilities (e.g. 3DTV) and increased quality. In the near future, 4K (or Ultra HD) will be available

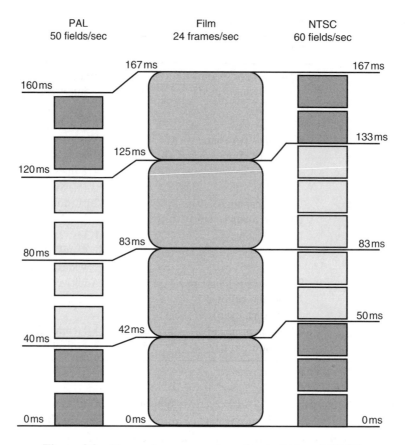

Figure 8.2   Conversion process from film to PAL and NTSC

increasing significantly the visual quality of the video sequences, providing extraor-dinarily vivid images and offering four times (or 16 times if it will be considered that it will not be interlaced) the resolution of a conventional HDTV.

8.3   Restoration of Video Noise and Grain

Any imaging system is effected by noise no matter how precise is the recording equip-ment. The sources of noise that can corrupt a film or a video sequence are numerous. Examples include the thermal or photonic noise in the television cameras and the film grain during the exposure and development of the film. Also the post-production may add significantly to the accumulated level of noise due to the nature of this process and the large number of the iterations it requires. Regarding grain, it is due to the depo-sition of particles on the film surface, which have been previously exposed to light. Grain is visible to areas of uniform colour and low image details, with the general rule to be that areas that receive less exposure to light to be grainer since the particles

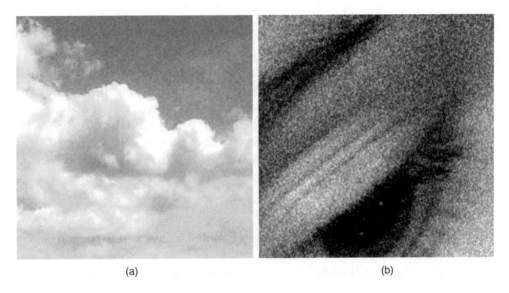

(a)   (b)

Figure 8.3 Examples of film grain (a) and image/video noise (b)

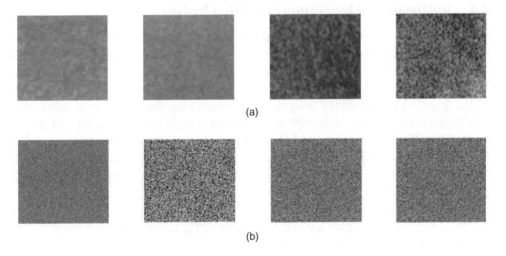

(a)

(b)

Figure 8.4 Examples of grain (top row) and video noise (bottom row) patterns

respond more to low levels of light. Examples of noise and grain frames are shown in Figure 8.3, while patterns of them are shown in Figure 8.4.

Image sequences stored in a digital form even if they are quite resistive to noise and these problems are alleviated, they cannot be totally eliminated [3–5]. Digital videos in general exhibit higher signal-to-noise ratio characteristics, since during the television production and the post-production the digital studios are capable of the highest quality standards. In digital video the noise can be added both during the acquisition and the transmission process due to pixel faults, jitter, lossy compression, loss of signal, environmental issues and so on.

The similarities of video noise and grain can be observed in Figure 8.3, but the human viewers are able to distinguish them. Noise in general lacks of a structure and appears spectrally flat, in comparison to film grain that appears in clusters and with a coarser structure. Despite these differences, during the restoration process both video noise and film grain are called noise and are approximated as additive noise with near-flat spectra, while the same mechanisms are used to alleviate their effects.

In practice, the effect of noise is modelled as an additive usually Gaussian noise with zero mean and variance σ^2 that is independent of the original uncorrupted video sequence $f(n, k)$. The corrupted image sequence $g(n, k)$ with the additive noise $w(n, k)$ is given by

$$g(n, k) = f(n, k) + w(n, k) \tag{8.1}$$

where $n = (n_1, n_2)$ corresponds to the special coordinates and k to the frame number of the sequence. The objective of restoration is to estimate an image sequence $\tilde{f}(n, k)$ that is as close as possible to the original given only the noise image sequence $g(n, k)$.

8.4   Restoration Algorithms for Video Noise

During the last four decades, noise reduction for digital images and video sequences is a subject of extensive research, mainly for military and satellite applications. Other applications include medical imaging, remote sensing and video broadcasting. Most of the algorithms developed are based on low-pass filtering, for example averaging, which is an intuitive first step towards image smoothing and hence noise reduction. The disadvantage of this approach is its effect on high frequencies (i.e. edges) of an image or frame generating a blurring distortion in the special domain and at the boarders of the moving objects in the time domain. Considerable amount of research was focused on motion estimation and registration, allowing the adaptation of filtering preventing undesirable artefacts in the temporal domain [6, 8, 9].

Let $g(n, k)$ be a discrete signal corrupted by noise and $h(l)$ a discrete filter carrying out a weighted averaging applied on successive signal samples (e.g. frames). The restored signal (image sequence) is given by

$$\tilde{f}_s(n, k) = \sum_{l=-K}^{K} h(l) g(n - l, k) \tag{8.2}$$

where $l = (l_1, l_2)$ the special coordinates of the filter. The boundaries of the summation define the length or the aperture of the filter, and this procedure is otherwise known as convolution. Typically, the larger the length of the filter, the more efficient is the filtering process but the computational complexity increases significantly and delays are introduced. Especially for images and all 2D signals, filtering is applied both on rows and columns, and therefore, the shape of the filter is considered during the design process. For video sequences, filtering can be in three dimensions considering also

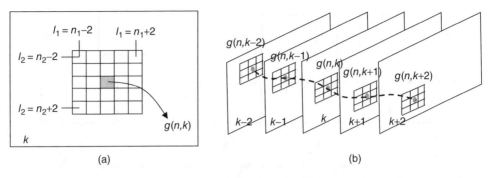

Figure 8.5 (a) Example of 5×5 special filter, (b) noise filter operating on the temporal axes

the temporal axes. In order to reduce the complexity of the filtering system, filters are designed to be variable-separable allowing the operation to be applied on one dimension at a time in a cascaded order. An example of a special filter using a 5×5 window is shown in Figure 8.5(a).

Designing a noise reduction filter operating in the temporal domain requires the image sequence pixels $g(n, k)$ and the filter coefficient elements $h(l)$. In this case, the pixel values are multiplied with the filter over time, and the restored frame is given by

$$\tilde{f}_t(n, k) = \sum_{l=-K}^{K} h(l)g(n, k - l) \tag{8.3}$$

The filter coefficients are considered equally important, but they can be optimized in a minimum mean-square error manner.

$$h(l) \leftarrow \min_{h(l)} E[(f(n, k) - \tilde{f}_t(n, k))^2] \tag{8.4}$$

Furthermore, filter coefficients should collectively return unity gain at zero frequency, that is $\sum_{l=-K}^{K} h(l) = 1$, and a simplified view of the arrangement is shown in Figure 8.6. This arrangement requires a lot of memory and therefore delays are introduced. Also, its performance depends on the image sequence contents, and in static scenes the accuracy is higher compared to the moving edges of the objects.

Instead of using the current and the past input, a recursive arrangement could be applied using a weighted average of past output frames

$$\tilde{f}_t(n, k) = \sum_{l=-K}^{K} h(l)\tilde{f}_t(n, k - l) + h(0) \tilde{f}_t(n, 0) \tag{8.5}$$

The recursive arrangement provides similar results, but utilizing fewer coefficients and a simple example of a first-order recursive filter is shown in Figure 8.7.

These filters as it was already mentioned are not suitable in areas of motion; therefore, these areas are estimated using frame registration. Motion estimation techniques

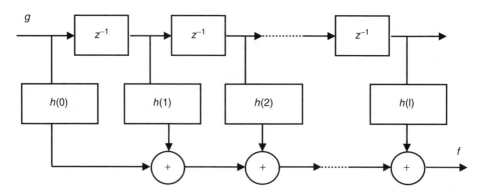

Figure 8.6   The arrangement of the temporal filtering system

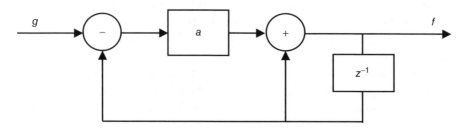

Figure 8.7   A first-order recursive filter, where $0 < a < 1$ is a scaling factor

could be utilized to estimate the scene movement and a threshold to separate the actual motion from the noise is applied, based on the assumption that significant frame differences are due to motion compared to noise. This approach can be further improved by taking into consideration that motion is also spatially correlated. The motion detector needs to operate on a pixel level to provide the motion trajectories of the picture elements. Therefore, either optical flow or block-based methods are required, with the first being more computational expensive and the second less accurate in terms of actual motion estimation especially for small block sizes. In order to improve the accuracy and reduce the complexity, either hierarchical or overlapping approaches can be incorporated to estimate the motion parameters.

The video restoration and the noise reduction performance can be improved if motion compensation is first applied to the previous frame prior computing the frame difference. So instead of operating on frame differences and determining the threshold to separate motion from noise there, the filtering operates on motion-compensated frame differences. Since the motion vectors are available, the filtering process can be applied along the picture elements that lie on the same motion trajectory. The restored sequence in this case is given by

$$\tilde{f}(n,k) = \sum_{l=-K}^{K} h(l)g(n_1 - \partial x(n_1,n_2;k,l), n_2 - \partial y(n_1,n_2;k,l), k-l) \qquad (8.6)$$

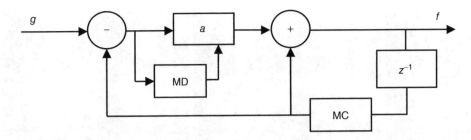

Figure 8.8   An example of motion-compensated filtering system.

where $\partial(n; k, l) = (\partial x(n_1, n_2; k, l), \partial y(n_1, n_2; k, l))$ is the motion vector for special coordinates (n_1, n_2) estimated between the frames k and l. A simple example of a motion-compensated filtering system is shown in Figure 8.8, with the motion compensation function applied on the previous output.

Another approach that is used in noise reduction is wavelet filtering, which is based on the observation that individual subbands contain a pre-specified portion of the spatial frequency spectrum of the original frame. Noise components especially in areas with low level of detail will not co-exist with actual image details in higher frequencies. A simple mechanism for wavelet-based filtering is to utilize zero-tree structures to associate coefficients at a lower frequency subband (parent) with co-sided coefficients at higher frequencies (children). In practice, if an insignificant parent has a significant child that indicates a potential presence of noise based on the principle that insignificant parents have on average insignificant children and vice versa.

Statistical approaches can also be applied to suppress visible noise based on noise modelling. A model is assumed for the existed noise, for example additive white Gaussian noise, and then a minimization process is applied to estimate the parameters of the models from the frame contents using mean-square error as a metric.

8.5   Instability Correction Using Registration

One of the most severe artefacts in a video sequence is regarded instability (or unsteadiness). The main causes of unsteadiness are related to either mechanical issues of the acquisition and transfer devices or the conditions during the capturing. For example, a camera on a moving vehicle could introduce unsteadiness reducing the quality of the film or video sequence (see Figures 8.9 and 8.10).

The artefact of instability is modelled mainly as a global translation from one frame to the next.

$$f(x, y, t) = f(x + u, y + v, t - 1) \tag{8.7}$$

In case of unsteadiness, this global translation $(u(t), v(t))$ is perceived as horizontal weaving and vertical hopping motions. These contain both random and periodic components based on the nature of the cause of the artefact.

Figure 8.9   Example frames of the Joan sequence indicating the motion of the camera

Figure 8.10   A close example of the Joan sequence comparing pairs of successive frames showing the global motion

The model of unsteadiness based on global translation can be further improved by taking into account fluctuations in the spatial domain $(u(x, y, t), v(x, y, t))$. This advanced model can be used in cases of geometric distortions due to faults in optics, and these artefacts are visible as spatial fluctuations across the image plane and mainly at the boarders. These models can be, for example, low-order polynomial approximations of these spatial distortions. The complexity of the model depends both on the source material and the viewers' quality experience. In general, unsteadiness is more noticeable is still frames, because in case of high camera and scene motion, the artefact of unsteadiness is partially masked.

The restoration process for unsteadiness is based on image registration, estimating the global translation parameters $(u(t), v(t))$. Motion estimation techniques such as block matching and phase correlation are suitable for this task. Since the motion parameters are obtained, compensation of the current frame t with respect to the previous frame $t-1$ is applied.

In block matching as it was discussed earlier, the entire frame is regarded and the best match on the basis of error minimization is computed. The best match is computed within search window in the previous frame as follows:

$$\arg \min_{u,v \in Z} \sum_i \sum_j |f_t(i,j) - f_{t-1}(i+u, j+v)| \tag{8.8}$$

where $f_t(i,j)$ is the (i,j)th pixel element at the frame t, (u,v) are the motion parameters that minimize the error and Z is the search window. Since the entire frame was selected, the solution of the above minimization problem results the global translation vector.

In case of phase correlation, the location of the maximum of the obtained correlation surface C_s corresponds to the global motion parameters.

$$\arg \max C_s(x, y) \tag{8.9}$$

where $C_s(x, y) = F^{-1}(F(f_{t-1}) * F(f_t)/|F(f_{t-1}) * F(f_t)|)$, F and F^{-1} denote forward and inverse 2D Fourier transform and $*$ denotes complex conjugate.

In order to improve the accuracy, more advanced models and approaches can be introduced considering the spatial characteristics of the frame motion. In practice, instead of registering the entire pictures and having a single motion vector, the frames are partitioned into smaller co-sided blocks typically of size 16×16 or 32×32 pixels. Then the same procedure is applied in each block and a motion vector is obtain per block using either block matching or phase correlation. The number of the obtained motion vectors in this case will be equal to the number of blocks in the frame, and the extraction of a single estimate can be done, for example, by taking the mean, the median and a least square approximation. These approaches can be further improved by taking into account an associated weight to each motion vector. This weight based on the confidence of the measurements (e.g. mean-square error and the maximum height of the correlation surface) and in case of mean is given by

$$(u_m, v_m) = 1/N \sum_{i=1}^{N} (w_i u_i, w_i v_i) \tag{8.10}$$

where N is the total number of blocks in a frame and w_i is the weight associated with the ith block. A more advanced model that incorporates also the spatial variations of unsteadiness can be obtained using an affine or polynomial approximation based on a least square fit.

The main issues of these algorithms are related to shot change and error propagation, therefore a fully automated restoration framework should consider mechanisms to detect shot changes and avoid the introduced error to propagate. Since a scene change is detected, the system will reset and the first frame of the new scene will be used as a reference frame for future unsteadiness correction. In this way, also the error propagation is reduced and a more robust restoration approach is provided. Many algorithms for automatic shot change detection are available in the literature, and it is worth also mentioning the work in [10, 11].

8.6 Estimating and Removing Flickering

A common artefact in video sequences is related to temporal frame intensity fluctuations, and it is called flickering. This artefact is caused mainly due to inconsistent exposure and environmental parameters. Particularly in films, multiple copies, errors in printing and ageing could be included in the causes of flickering. Flickering is a well-known artefact and immediately recognizable in old sequences even by non-expert viewers. Its perceptual impact can be significant, and it can lead to considerable discomfort and eye fatigue after extended viewing. Flickering is mainly noticeable in static scene or in frames with low activity since camera and content motion can partially mask its effect.

All the artefacts discussed earlier such as noise and unsteadiness including flickering belong to the same category of global artefacts as they usually effect the entire frame as opposed to so-called local artefacts such as dirt, dust or scratches, which are usually present to specific locations on the frames [12–17].

In order to provide restoration solutions to flicker, a model of this artefact is required. A simple approach is to consider flickering as a global intensity shift from one frame to the next, which is equivalent to adding a constant to each pixel and only affects the mean intensity.

$$f(x, y, t) = f(x, y, t - 1) + c(t) \tag{8.11}$$

Another model that is commonly used for flickering considers it as a global intensity multiplier, which models also fluctuations in intensity variance.

$$f(x, y, t) = a(t)f(x, y, t - 1) \tag{8.12}$$

A more flexible approach combining both of the above models was introduced regarding the parameters $a(t)$ and $c(t)$ as random processes.

$$f(x, y, t) = a(t)f(x, y, t - 1) + c(t) \tag{8.13}$$

Flickering can also be modelled independently of the previous frames as a linear combination of the unaffected original one plus noise.

$$\tilde{f}(x, y, t) = a(t)f(x, y, t) + c(t) + n(t) \tag{8.14}$$

This approach allows considering flickering as a result of noise independent on time. Furthermore, it can be improved by considering also special variations simulating varying intensity fluctuations over the frame plane, which depend on the scene contents.

The restoration of flicker in a video sequence can improve significantly the viewing experience and enhance the temporal and special scene details. A general framework for flickering correction consists of the following steps.

Initially, the parameters $c(t)$ and $a(t)$ of the above models are estimated. The global intensity shift $c(t)$ is obtained by subtracting the mean intensities of two successive frames as we can see below.

$$c(t) = \bar{f}(x, y, t) - \bar{f}(x, y, t - 1) \tag{8.15}$$

While the global intensity multiplier $a(t)$ is given by the ratio of the equivalent frame mean intensities.

$$a(t) = \bar{f}(x, y, t) / \bar{f}(x, y, t - 1) \tag{8.16}$$

where $\bar{f}(x, y, t)$ is the mean value of $f(x, y, t)$. Since these parameters are available, the inverse procedure is applied and the restored frames without flickering are obtained.

This approach is sensitive to local intensity variations due to scene and camera movement; therefore, it is used mainly for semi-automatic methods where a trained person selects an area of the frame without motion and the algorithm is applied only there. In order to improve the performance and provide a fully automated method even in the presence of motion, robust measurements and techniques were introduced. For example, median, histogram intensity maximization and motion compensation-based methods were proposed.

Significant improvement can be obtained in flickering restoration by incorporating motion estimation techniques. The idea is to detect the moving areas and exclude them from the process discussed earlier. Also it is crucial to keep the complexity low allowing real-time restoration. A simple approach to identify the areas is to calculate the absolute frame difference and apply a threshold to obtain a binary segmentation indicating the presence or the absence of motion.

$$S_D(x, y, t) = \begin{cases} 1 & \text{if } |f(x, y, t) - f(x, y, t - 1)| \geq T \\ 0 & \text{if } |f(x, y, t) - f(x, y, t - 1)| < T \end{cases} \tag{8.17}$$

The main disadvantage of this approach is the threshold selection, since if it is set too high small movements may be ignored while if it is too low flickering may be regarded as movement. Therefore, an approach that could adapt to the image contents could eliminate this problem, and a simple one can be based on absolute frame difference histograms. In this case, a fixed percentage of the cumulative distribution of the frame pixel intensities is selected as the threshold to separate the static areas in the scene [18–25].

Another approach that takes into consideration spatial variations of flickering is to split successive frames into blocks (e.g. 16×16 or 32×32 pixels) and apply the same process that was analysed above to each one. Then in order to obtain a single global estimate of the parameters of the flickering model, the mean or the median of the values of all the available blocks is calculated. Additionally, this method can be used to estimate an affine polynomial model allowing more accurate special adaptation.

In case a memory-less model is used, a linear minimum mean-square error estimator is considered. The only information that is required is the mean and variance of the noise in the current frame that is assumed to be provided from the noise statistics of the previous one. Based on this approach, the model parameters will be given by

$$a(t) = \sqrt{f_\sigma(x, y, t) - n_\sigma(x, y, t)/f_\sigma(x, y, t - 1)}$$
$$c(t) = f_\mu(x, y, t) - a(t)f_\mu(x, y, t - 1)) \tag{8.18}$$

where $f_\mu(x, y, t)$ and $f_\sigma(x, y, t)$ denote the mean and the variance of the frame f, respectively. Also the block-based approach can be applied in this case, as it was analysed earlier and with the option to have an affine model fitted. An example of a sequence with flickering and the restored version is shown in Figure 8.11.

Other methods that were proposed for flickering removal are based on filtering and interpolation applied iteratively, which are called successive over-relaxation

Figure 8.11 Example of stabilization of flicker-like effects in image sequences through local contrast correction

techniques. Also, histogram-based methods can be used to measure the global features of the frame providing estimates of the flickering model parameters. For example, the difference of the maximum values of the histograms or equivalently the maximum of the histogram of the frame difference of two successive frames can provide the parameter $a(t)$ of the model.

$$a(t) = \arg\max \ \{H(f(t)) - \arg\max H(f(t-1))\}$$

$$a(t) = \arg\max\{H(f(t) - f(t-1))\} \tag{8.19}$$

where H is the histogram of the corresponding frame or frame difference. Furthermore, flickering can be removed by working on a local level (e.g. pixel based) instead of considering it as a global artefact. The intensity of each of the frame pixels is estimated as a weighted average of the corresponding previous, current and future values:

$$\tilde{f}(x, y, t) = \sum_{t_s=-T}^{T} w(t)f(x, y, t + t_s) \tag{8.20}$$

where $w(t)$ is the filter weight. The performance of this approach depends of the filter length T and for larger T more accurate results are obtained. The drawback of larger filters is the requirement for more frames to be stored in memory introducing delays and increased computational complexity and cost. Therefore, based on the application and the level of tolerance, the approach can be modified to fit the required needs of the system. For example, only the past, frames may be considered to avoid delays and also spatial filtering could be introduced to allow more accurate and smooth flickering removal [15, 18, 26].

8.7 Dirt Removal in Video Sequences

Dirt and scratch reduce the quality and the pleasure of film watching, and additionally a significant amount of visual information is lost. Regarding the missing information, the human brain is very effective at substituting missing relevant information based on its prior experience of vision. For example, dirt deposited on the film during scanning or screening appears in one frame and disappears in the other. This contradicts with real-life scenes where the objects are moving smoothly and the dirt artefact causes the brain to treat this abnormal visual stimulus exceptionally. Therefore, dirt and scratch restoration aims to keep the viewers relax during the film allowing them to focus on the contents and the story [27–38].

Small particles on the film surface, the camera lenses or the projector can result scratches that appear in multiple frames almost in the same position destroying temporal information. Also scratches can have the shape of vertical lines, and some examples are shown in Figure 8.12. Dirt mainly appears on a single frame and has similar characteristics with a moving object. Therefore, it is regarded a hard problem to distinguish actual object motion from dirt especially in cases of erratic motion, motion blur and occlusion. Examples of dirt are shown in Figures 8.13 and 8.14.

Observing the images above with dirt and scratches, it can be inferred that these artefacts appear either as dark or bright areas, which depends on if the dirt particles

Figure 8.12   Example frames with scratches, vertical and non-vertical

Figure 8.13   An example of a degraded frame is shown at the left, while the ground truth is denoted with dark spots on the right image indicating the locations of dirt

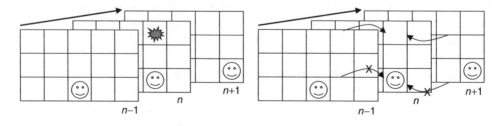

Figure 8.14   Example of dirt detection and removal

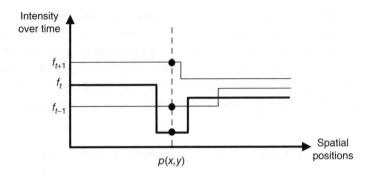

Figure 8.15   Example of intensity variation over time in locations with dirt artefacts

adhere to positive or negative film, respectively. The most common model for dirt is an impulse noise in the time domain, and also it is assumed that it will be present at a specific spatial location of a particular frame with low likelihood to appear at the same location in more than one successive frames.

A significant number of dirt restoration algorithms are based on the above model exploited in different ways, and most of them are two-step approaches including a detection and a removal part. One approach known as dirt signature detection requires three successive frames, the current one that is restored, the previous and the future frame. Assuming that at a certain location (x,y) of the current frame f_t dirt is present, then if we observe the intensity variations over time at this specific location, the variation over time is significantly higher than in areas without any artefacts (see Figure 8.15). In practice, dirt is detected if all the following conditions are simultaneously true:

$$|f_{t-1}(x, y) - f_t(x, y)| > T$$
$$|f_{t+1}(x, y) - f_t(x, y)| > T \quad\quad\quad (8.21)$$
$$\mathrm{sgn}(f_{t-1}(x, y) - f_t(x, y)) == \mathrm{sgn}(f_{t+1}(x, y) - f_t(x, y))$$

where sgn() returns the sign of its argument and T is a predefined threshold. This algorithm is very accurate in static areas or of low motion but since the video sequences contain scene and camera motion, a registration method is essential. The registration step will reduce the number of moving objects being erroneously detected as dirt. Also, unsteadiness may compromise the performance of dirt detection, and therefore it is essential to remove this artefact prior the dirt detection mechanism.

The registration process can be based on either advanced motion estimation algorithms [39, 40] or just the frame differences can be used. In this case, the absolute difference is considered, and a low-pass filter is applied on the result. The obtained estimate is threshold to define the presence or not of motion in the scene. Since the activity in each area of a frame is estimated, the above dirt signature detection algorithm is applied only in the areas with low motion.

Filtering-based approaches have been developed to remove dirt artefacts, and the most common approach is based on median filtering. The main characteristic of median filter is that it does not affect sharp edges and no blurring is introduced. Also, the shape and the size of the filter are part of the design process and are based on the application. Incorporating median filtering for dirt removal is required to expand it to three dimensions to include time. The median filter operation centred on a pixel $f(x, y, t)$ of a frame f is given by

$$\text{med}\{f(x + x_0, y + y_0, t + t_0)\} \quad \text{with}\{x_0, y_0, t_0\} \in W \qquad (8.22)$$

where W is the selected spatiotemporal window of pixels. Since this filtering approach is computational, expensive pseudo-median and separable-median mechanisms were proposed that do not yield exactly the same results but acceptable approximations.

Median filtering is regarded a good candidate mechanism for the removal of dirt due to the fact that this artefact is characterized by its short-duration discontinuities in the space-time domain. Also, it was been proved experimentally that temporal median filtering is more effective compared to the one in the spatial domain, but this depends on the scene contents and their movement. In sequences that contain fast motion, dirt may be confused with moving objects, and as a result the latter to be filtered out. Examples are the objects with speed higher than the frame rate, and as a result they appear as impulse signals in the time domain.

The removal of dirt can be significantly improved if the temporal median filter instead of operating on co-sided in consecutive frames; motion-compensated pixels are considered (see Figure 8.16).

$$\text{med}\{f(x + u, y + v, t + t_0)\} \quad \text{with} \quad t_0 \in W \qquad (8.23)$$

where (u, v) is the estimated motion vector at the pixel location (x, y). This implies that the method requires a dense motion field to be estimated for consecutive frames. A simple block diagram of this approach is shown in Figure 8.17.

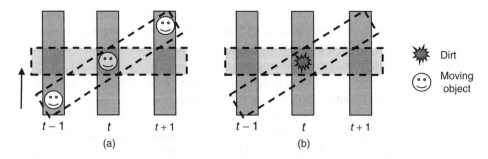

Figure 8.16 Median filtering on the co-sided pixels (dark coloured) and on the motion-compensated pixel locations (transparent) for both moving objects (left) and dirt artefacts (right)

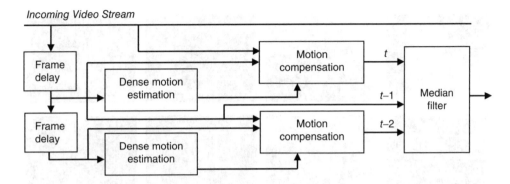

Figure 8.17   Block diagram of dirt removal using motion compensation

Dirt concealment using morphological operators is another well-known approach in digital film restoration. This is due to the fact that these morphological operators (e.g. opening and closing) can eliminate small structures such as dirt particles. The main disadvantage is that they result a smoothing effect, and they cannot remove efficiently large dirt artefacts. Another approach is to apply them in the temporal domain since the structure of dirt in this domain is more suitable for morphological operators. Also, robust or soft morphological operators can be utilized, which relax the standard definitions of these operators by introducing robust or general weighted order statistics [7, 40–47].

8.8   Metrics in Video Restoration

In video and film restoration methods to evaluate the performance of the algorithms and the quality of the restored sequences, metrics are widely used and are based on the same principles that were discussed in previous chapters. The evaluation process may be either quantitative or qualitative, providing different ways to determine the performance of restoration techniques and the users' satisfaction [8, 15, 48].

In order to apply a quantitative analysis, a certain procedure is required including evaluation datasets with ground truth and appropriate metrics selection. Restoration algorithms presented in this chapter can be tested on several real images affected by real artefacts such as noise, dirt, line scratches, unsteadiness and flickering. Examples of evaluation sequences could be the Knight, the Italian movie Animali che attraversano la strada, and the sequence Sitdown available in [33, 49, 50] that contain mainly scratches and dirt. Also, the broadcast resolution (720 × 576 pixels) sequence 'Pennine Way', which contains fast motion and textured background, is a common video used during the experiments. Other video sequences affected by real artefacts are the 'Lamp of Memory' greyscale image sequence with static background and slow movements of the object in the foreground and the 'Landmine Warfare', which was shot in 1962 containing a lot of bright points due to intensity flicker. In Figure 8.18, we can

Figure 8.18   Examples of video and film sequences used in restoration for evaluation

see examples of different sequences (e.g. Kiss, Chaplin, Woman, Sitdown, Man and Woman, Thalamus, Garden and Frank) used in restoration containing dirt, flickering or other artefacts.

A corresponding ideal (artefact-free) reference sequence is required in order to carry out an evaluation using quantitative metrics such as the mean absolute error (MAE) criterion [51]. In practice such a reference is unavailable; therefore, in most of the evaluation processes, manually derived ground truth of dirt, flickering, scratches, noise and unsteadiness are employed for quantitative evaluations. The approach that is used to induce artificial artefacts in a video sequence in order to obtain ground truth varies based on the particular artefact. For example in case of noise, Gaussian additive noise is superimposed on the original sequence; unsteadiness is modelled as sharp global translations applied on the original frames; flickering could be obtained by applying the model in Equation (8.14) altering the global frame intensity; and regarding scratches and dirt, binary marks are digitally superimposed on the film providing sequences with artefacts and ground truth for quantitative analysis.

Based on manually generated ground truth, a quantitative performance assessment can be carried out using different key criteria and metrics. One of the most common metrics that is widely used is the correct detection rate R_c, defined as follows:

$$R_c = \text{Count}(D_g \text{ AND } D_e)/\text{Count}(D_g) \tag{8.24}$$

where D_g is the ground truth mask and D_e is the detection mask obtained from any given method. Count is a function counting the non-zero elements in a mask, and operator AND is the logical AND defined as follows:

$$(D_a \text{ AND } D_b)(i,j) = \begin{cases} 1 & \text{if } D_a(i,j) * D_b(i,j) \neq 0 \\ 0 & \text{otherwise} \end{cases} \qquad (8.25)$$

Other criteria could be the false alarm rate R_f and the missed detection rate R_m that are based on the same concept as correct detection rate in Equation (8.24) and defined as follows:

$$R_f = \text{Count}(D_e \text{ AND } !D_g)/\text{Count}(D_g) \qquad (8.26)$$

and

$$R_m = \text{Count}(D_g \text{ AND } !D_e)/\text{Count}(D_g) = 1 - R_c \qquad (8.27)$$

where $!D_a(i,j) = \begin{cases} 1 & \text{if } D_a(i,j) = 0 \\ 0 & \text{otherwise} \end{cases}$. The range of these criteria is determined as $0 \leq R_c, R_m \leq 1$ and $0 \leq R_f \leq N_o/N_g - 1$, where N_o and N_g are the number of pixels in the original frame image and the number of artefact pixels in D_g, respectively. These metrics are mainly applicable when a mask representing the ground truth is available; therefore, they can be utilized in artefacts such as dirt, scratches and is some type of additive noise.

Furthermore, in order to obtain objective measures of the reconstruction algorithms accuracy, their results are compared with the artificially corrupted images I_c with the ideal uncorrupted images. Specifically, the original image I_o and the restored image I_r obtained with any one of the considered removal algorithms are utilized in the following objective measures:

$$\text{MSE} = \frac{1}{M * N} \sum \sum \| I_o - I_r \|^2 \qquad (8.28)$$

which provides the MSE between the original and the restored images, with $\|.\|$ denoting the vector norm and the parameters M, N the frame size. Such measure gives a non-negative value; the smaller the value of MSE, the better the restoration result.

$$\text{PSNR} = \frac{1}{M * N} \sum \sum \left(10 * \log_{10} \left(\frac{255^2}{\|I_o - I_r\|^2} \right) \right) \qquad (8.29)$$

The peak signal-to-noise ratio between the original and the restored images obtained considering the MSE. Such measure gives a non-negative value; the higher the value of PSNR, the better the restoration result.

$$\text{SSIM} = \frac{1}{M * N} \sum \sum \frac{(2 * \mu I_o * \mu I_r + C_1)(2\sigma I_o I_r + C_2)}{(\mu I o^2 + \mu I_r^2 + C_1)(\sigma I o^2 + \sigma I_r^2 + C_2)} \qquad (8.30)$$

Another measure is the structural similarity index [52] that is applied to the original and the restored images, where $C_1 = (K_1 * L)^2, C_2 = (K_2 * L)^2, K_1 = 0.01, K_2 = 0.03$ and $L = 255$. Such a metric gives values in $[0, 1]$; the higher the value of SSIM, the better the restoration result.

Second-order statistic and image entropy can also be used to evaluate the results of the restoration process [53, 54]. Second-order statistic [55] is defined as

$$\text{SOS} = \frac{1}{M * N}(\sum I_r^2(i,j)(\sum I_r(i,j))^2) \tag{8.31}$$

The larger the second-order statistic is, the better the result of image restoration will be. The image entropy [56] is defined as follows:

$$H = -\sum_{i=0}^{L-1} p_i \log p_i \tag{8.32}$$

where L is the whole image grey scale, p_i represents the rate of pixels D_k whose grey scale is k, whole image pixel is D, defined as

$$p_i = D_k/D \tag{8.33}$$

and based on that the image has more information when its entropy is larger. In specific cases such as in flickering and unsteadiness, the variance of the intensity fluctuations and the global translation differences could be considered as a measure, respectively.

Regarding the complexity of the restoration algorithms, experimental execution times with corresponding estimates are compared. During this evaluation process, experimental execution times are captured without including input and output of video sequences. It should be mentioned that all the experiments should be performed under the same conditions in terms of both the available hardware/software and the input data. Furthermore, the complexity of the methods can be measured using the number of multiplications and summations if that can be calculated from either the restoration model or the implemented algorithm.

Qualitative metrics could also be utilized to evaluate the performance of film and video restoration algorithms, in a similar manner as in the case of video coding. A dataset of restored videos with different types of artefacts and severity level are observed by a group of humans (including both experts and non-experts), providing a feedback on each sequence. The restored sequences with the higher quality as it was obtained by the evaluation process will indicate the most accurate restoration method. Additionally, research on designing quantitative metrics that are based on qualitative evaluations is conducted as it was discussed in Chapter 2, and these measures are based on the assumption that the level of distortion is known (e.g. artificially induced). The evaluation process is similar to the qualitative one, and the most accurate quantitative metric is considered the one that is closer to the outcomes of the qualitative analysis.

8.9   Conclusions

In this chapter, motion estimation techniques used in video restoration are discussed and analysed. Initially, an overview of video and film restoration approaches is provided, focusing on their similarities and differences. Also, the evolution of the techniques and technologies involved in the restoration process of videos and archive sequences is presented. The main types of degradation and common artefacts in videos and films are presented, and their general causes are analysed. Restoration methods based on registration for noise and grain are discussed. Solutions to issues such as instability and flickering are presented operating in either the pixel or the frequency domain. All the types of dirt in film sequences are analysed, providing different approaches to overcome this issue using global and local registration techniques. Finally, common evaluation methods and metrics used in video restoration are discussed, while available datasets or public sequences are presented that are utilized in state-of-the-art comparative studies.

8.10   Exercise–Practice

This exercise is focused on instability correction using global registration. Using the video sequence Joan, shown also in Figure 8.9, remove the unsteadiness using global motion estimation (e.g. in the frequency or the pixel domain) and image compensation.

In order to remove this instability, a two-step approach is needed. First using, for example, phase correlation estimate the global motion for each pair of frames. To obtain the global motion vector, simply apply phase correlation on the whole frame. Since the global motion is available, motion compensation should be applied to remove the unsteadiness, resulting a new video sequence without this artefact.

References

[1] Kokaram, A. (2007) Ten years of digital visual restoration systems. *IEEE International Conference on Image Processing, September 2007. ICIP 2007*, Vol. **4**, pp. IV–1–IV–4.
[2] Ceccarelli, M. and Petrosino, A. (2002) High performance motion analysis for video restoration. *2002 14th International Conference on Digital Signal Processing, 2002. DSP 2002*, Vol. **2**, pp. 689–692.
[3] Chan, S., Khoshabeh, R., Gibson, K. *et al.* (2011) An augmented lagrangian method for video restoration. *2011 IEEE International Conference on Acoustics, Speech and Signal Processing (ICASSP), May 2011*, pp. 941–944.
[4] Chang, R.-C., Tang, N. and Chao, C.-C. (2008) Application of inpainting technology to video restoration. *2008 First IEEE International Conference on Ubi-Media Computing, July 2008*, pp. 359–364.
[5] Chan, S., Khoshabeh, R., Gibson, K. *et al.* (2011) An augmented lagrangian method for total variation video restoration. *IEEE Transactions on Image Processing*, **20** (11), 3097–3111.
[6] Richardson, P. and Suter, D. (1995) Restoration of historic film for digital compression: a case study. *Proceedings, International Conference on Image Processing, October 1995*, Vol. **2**, pp. 49–52.

[7] Wu, Y. and Suter, D. (1995) Historical film processing. *Proceedings of SPIE*, **2564**, 289–300.

[8] Schallauer, P., Bailer, W., Morzinger, R. *et al.* (2007) Automatic quality analysis for film and video restoration. *IEEE International Conference on Image Processing, September 2007. ICIP 2007*, Vol. 4, pp. IV–9–IV–12.

[9] Harvey, N. and Marshall, S. (1998) Video and film restoration using mathematical morphology. *IEE Colloquium on Non-Linear Signal and Image Processing (Ref. No. 1998/284), May 1998*, pp. 3/1–3/5.

[10] Kim, W.-H., Moon, K.-S. and Kim, J.-N. (2009) An automatic shot change detection algorithm using weighting variance and histogram variation. *11th International Conference on Advanced Communication Technology, February 2009. ICACT 2009*, Vol. **2**, pp. 1282–1285.

[11] Yao, S.-W. and Liu, J.-W. (2007) Automatic scene change detection for H.264 video coding. *International Conference on Wavelet Analysis and Pattern Recognition, November 2007. ICWAPR '07*, Vol. **1**, pp. 409–414.

[12] Delon, J. (2006) Movie and video scale-time equalization application to flicker reduction. *IEEE Transactions on Image Processing*, **15** (1), 241–248.

[13] Pitie, F., Dahyot, R., Kelly, F. and Kokaram, A. (2004) A new robust technique for stabilizing brightness fluctuations in image sequences. *2nd Workshop on Statistical Methods in Video Processing in conjunction with ECCV*.

[14] Kokaram, A., Dahyot, R., Pitie, F. and Denman, H. (2003) Simultaneous luminance and position stabilization for film and video. *Proceedings of SPIE 5022, Image and Video Communications and Processing*.

[15] Schallauer, P., Pinz, A. and Haas, W. (1999) Automatic restoration algorithms for 35mm film. *Journal of Computer Vision Research*, **1** (3), 60–85.

[16] Naranjo, V. and Albiol, A. (2000) Flicker reduction in old films. *IEEE Proceedings of ICIP*, Vol. **2**, pp. 657–659.

[17] Roosmalen, P., Lagendijk, R. and Biemond, J. (1999) Correction of intensity flicker in old film sequences. *IEEE Transactions on Circuits and Systems for Video Technology*, **9** (7), 1013–1019.

[18] Forbin, G. and Vlachos, T. (2008) Nonlinear flicker compensation for archived film sequences using motion-compensated graylevel tracing. *IEEE Transactions on Circuits and Systems for Video Technology*, **18** (6), 803–816.

[19] Vlachos, T. (2004) Flicker correction for archived film sequences using a nonlinear model. *IEEE Transactions on Circuits and Systems for Video Technology*, **14** (4), 508–516.

[20] Vlachos, T. (2001) A non-linear model of flicker in archived film sequences. *Proceedings of IEE SEM on Digital Restoration of Film and Video Archives*, 01/049:10.1–10.7.

[21] Vlachos, T. and Thomas, G.A. (1996) Motion estimation for the correction of twin-lens telecine flicker. *IEEE Proceedings of ICIP*, Vol. **1**, pp. 109–112.

[22] Forbin, G. and Tredwell, T.V.S. (2006) Spatially-adative non-linear correction of flicker using greylevel tracing. *Proceedings of the 7th Symposium on Telecommunications, Networking and Broadcasting*.

[23] Forbin, G. and Vlachos, T. (2007) Flicker compensation for archived film using a mixed segmentation/block-based nonlinear model. *Proceedings of EUSIPCO*.

[24] Forbin, G., Vlachos, T. and Tredwell, S. (2006) Flicker compensation for archived film using a spatially-adaptive nonlinear model. *IEEE Proceedings of ICASSP*, Vol. **2**, pp. 77–80.

[25] Forbin, G., Vlachos, T. and Tredwell, S. (2005) Spatially adaptive flicker compensation for archived film sequences using a nonlinear model. IEE Proceedings of CVMP, pp. 241–250.

[26] Forbin, G. and Vlachos, T. (2008) Flicker compensation for archived film sequences using a segmentation-based nonlinear model. *EURASIP Journal on Advances in Signal Processing*, **2008**, 347495, 1–16.

[27] Gai, J. and Kang, S. (2009) Matte-based restoration of vintage video. *IEEE Transactions on Image Processing*, **18** (10), 2185–2197.

[28] Corrigan, D., Harte, N. and Kokaram, A. (2008) Pathological motion detection for robust missing date treatment. *EURASIP Journal on Advances in Signal Processing*, **2008**, 542436, 1–16.

[29] Wang, X. and Mirmehdi, M. (2009) HMM based archive film defect detection with spatial and temporal constraints. *Proceedings of the British Machine Vision Confeernce (BMVC?09)*.

[30] Sorin, T., Isabelle, B. and Louis, L. (2007) Fusion of complementary detectors for improving blotch detection in digitized films. *Pattern Recognition Letters*, **28** (13), 1735–1746.

[31] Abe, S.N.M. and Kawamata, M. (2008) Fast and efficient MRF-based detection algorithm of missing data in degraded image sequences. *IEICE Transactions on Fundamentals of Electronics, Communications and Computer Sciences*, **E91-A** (8), 1898–1906.

[32] Licsar, A., Szirnyi, T. and Czni, L. (2010) Trainable blotch detection on high resolution archive films minimizing the human interaction. *Machine Vision and Applications*, **21** (5), 767–777.

[33] Maddalena, L., Petrosino, A. and Laccetti, G. (2009) A fusion-based approach to digital movie restoration. *Pattern Recognition*, **42** (7), 1485–1495.

[34] Maddalena, L. and Petrosino, A. (2008) Restoration of blue scratches in digital image sequences. *Image and Vision Computing*, **26** (10), 1314–1326.

[35] Kim, K.-T. and Kim, E.Y. (2010) Film line scratch detection using texture and shape information. *Pattern Recognition Letters*, **31** (3), 250–258.

[36] Tilie, S., Laborelli, L. and Bloch, I. (2006) Blotch detection for digital archives restoration based on the fusion of spatial and temporal detectors. *9th International Conference on Information Fusion, July 2006*, pp. 1–8.

[37] Maragos, P. (2005) Morphological filtering for image enhancement and feature detection, in *The Image and Video Processing Handbook*, 2nd edn (ed. A.C. Bovik), Elsevier Acad. Press, pp. 135–156.

[38] Gangal, A., Kayikioglu, T. and Dizdaroglu, B. (2004) An improved motion-compensated restoration method for damaged color motion picture films. *Signal Processing: Image Communication*, **19** (4), 353–368.

[39] Zitova, B. and Flusser, J. (2003) Image registration methods: a survey. *Image and Vision Computing*, **21**, 977–1000.

[40] Ren, J. and Vlachos, T. (2010) Detection of dirt impairments from archived film sequences: survey and evaluations. *SPIE Optical Engineering*, **49** (6), 1–12.

[41] Ren, J. and Vlachos, T. (2007) Missing-data recovery from dirt sparkles on degraded color films. *SPIE Optical Engineering*, **46** (7), 0770011.

[42] Ren, J. and Vlachos, T. (2007) Segmentation-assisted detection of dirt impairments in archived film sequences. *IEEE Transactions on Systems, Man, and Cybernetics Part B: Cybernetics*, **37** (2), 463–470.

[43] Ren, J. and Vlachos, T. (2007) Efficient detection of temporally impulsive dirt impairments in archived films. *Signal Processing*, **87** (3), 541–551.

[44] Ren, J. and Vlachos, T. (2005) Segmentation-assisted dirt detection for archive film restoration. *Proceedings of BMVC*.

[45] Ren, J. and Vlachos, T. (2007) Detection and recovery of film dirt for archive restoration applications. *IEEE Proceedings of ICIP*.

[46] Ren, J. and Vlachos, T. (2005) Dirt detection for archive film restoration using an adaptive spatio-temporal approach. *IEE Proceedings of CVMP*, pp. 219–228.

[47] Ren, J. and Vlachos, T. (2006) Non-motion-compensated region-based dirt detection for film archive restoration. *SPIE Optical Engineering*, **45** (8), 1–8.

[48] Schallauer, P. and Morzinger, R. (2006) Rapid and reliable detection of film grain noise. *2006 IEEE International Conference on Image Processing, October 2006*, pp. 413–416.

[49] Kokaram, A.C. (1998) *Motion Picture Restoration: Digital Algorithms for Artefact Suppression in Degraded Motion Picture Film and Video*, 1st edn, Springer-Verlag, London.

[50] DIAMANTProject (2011) (ecist 199912078) digital film manipulation system, http://diamant .joanneum.ac.at/ (accessed 14 February 2015).

[51] Hamid, M., Harvey, N. and Marshall, S. (2003) Genetic algorithm optimization of multidimensional grayscale soft morphological filters with applications in film archive restoration. *IEEE Transactions on Circuits and Systems for Video Technology*, **13** (5), 406–416.

[52] Wang, Z., Lu, L. and Bovik, A. (2004) Video quality assessment based on structural distortion measurement. *Signal Processing: Image Communication*, **19**, 121–132.

[53] Xiaoli, Z. and Bhanu, B. (2007) Integrating face and gait for human recognition at a distance in video. *IEEE Transactions on Systems, Man, and Cybernetics Part B: Cybernetics*, **37** (5), 1119–1137.

[54] Yuan, J.-H., YIN, X.-M. and ZOU, M.-Y. (2006) A fast super-resolution reconstruction algorithm. *Journal of the Graduate School of the Chinese Academey of Sciences*, **23** (4), 514–519.

[55] Pinson, M. and Wolf, S. (2004) A new standardized method for obejectively measuring video quality. *IEEE Transactions on Broadcasting*, **50** (5), 312–322.

[56] Periaswamy, S. and Farid, H. (2003) Elastic registration in the presence of intensity variantions. *IEEE Transactions on Medical Imaging*, **22**, 865–874.

Index

Image, Video & 3D Data Registration: Medical, Satellite & Video Processing Applications with Quality Metrics,
First Edition. Vasileios Argyriou, Jesus Martinez Del Rincon, Barbara Villarini and Alexis Roche.
© 2015 John Wiley & Sons, Ltd. Published 2015 by John Wiley & Sons, Ltd.